Public and Media Relations for the Fire Service

Public and Media Relations for the Fire Service

Tim Birr

FIRE ENGINEERING®

PennWell
MEDIA FOR STRATEGIC MARKETS SINCE 1910

Published by Fire Engineering Books & Videos
A Division of PennWell Publishing Company
Park 80 West, Plaza 2
Saddle Brook, NJ 07663
United States of America

James J. Bacon, editor
Book design: Max Design
Cover design: Steve Hetzel

Printed in the United States of America

1 2 3 4 5 6 7 8 9 10

Library of Congress Cataloging-in-Publication Data

Birr, Tim, 1953-
      Public and media relations for the fire service / Tim Birr.
            p.   cm.
      ISBN 0-912212-79-9 (softcover)
      1. Fire departments—Public relations. I. Title.
TH9158.B57   1999
659.2'936337—dc21   98-48123
                                                     CIP

About the Author

Tim Birr began his fire service career in 1975 as a career firefighter-EMT with the Eugene (OR) Fire Department and spent more than nine years there as a line firefighter, lieutenant, and captain. On a part-time basis, he began serving as the department's first public information officer in 1979 and went on to serve nearly a year as acting public information director for the City of Eugene. Ultimately he served for a decade as the primary spokesman and media liaison for police, fire, EMS, 911, and emergency management.

In 1995, Birr joined Tualatin Valley Fire and Rescue, Oregon's largest fire district, where he holds the rank of division chief and manages media, community, and intergovernmental relations.

In addition to his primary work, he has served as vice-president of IAFF Local 851, a media consultant for the U.S. Fire Administration, and state editor of the *Oregon Professional Fire Fighter* magazine. His freelance writings have appeared in numerous fire and general interest publications, including a "My Turn" column in *Newsweek*. For a year, he hosted a television talk show on health and safety issues for a local Group W affiliate.

A former regional director of the National Information Officers Association, Birr has instructed and lectured extensively on the topic of media relations to fire service and other emergency responders.

Dedication

For Meghan, Brendan, and the firefighters of Eugene and Tualatin Valley, past, present, and future.

Acknowledgments

There are a lot of people who contributed directly or indirectly to the writing of this book, and the list of those who've helped me over the course of nearly two decades is too long to include here. There are a number of people who do deserve specific thanks, including Kimberly Heilman-Sobie, J.D., LL.M., for her gracious assistance in researching the legal cases included in Chapter Five; Phil Lemman, for his counsel on public records law and political matters; Tualatin Valley Fire and Rescue Chief Jeff Johnson, for his insights into marketing and customer service; and Frank Cowan of the California Specialized Training Institute, for continuous information sharing. A special thanks goes to Paul LeSage and Seth Walker, who both at various stages helped a computer-phobic author with the intricacies of high-tech word processing. Diane Feldman of Fire Engineering Books and Videos deserves thanks for her patience and editing skill.

You always learn from the people you work with, so my acknowledgments to partners past and present include thanks to Tim McCarthy, Jan Power, Karen Eubanks, and Kristin Chaffee.

Afraid to leave someone out, I'll thank in general terms

the faculty of the University of Oregon School of
Journalism and the gang at the National Association of
Information Officers, who have contributed to my profes-
sional education in manifold ways.

Finally, my deepest appreciation goes to those journalists with whom I've had the pleasure of working over the
years, especially those who've been kind enough to share
off-duty evenings discussing the practice, ethics, and
stresses of their important craft. Despite professionally
adversarial relationships, I number many of them among
my best friends.

Photo Credits

About the Author: J. David Straub
Chapter One: Bill Stormont
Chapter Two: Portland Fire Bureau
Chapter Three: Portland Fire Bureau
Chapter Four: Portland Fire Bureau
Chapter Five: Bill Stormont
Chapter Six: Portland Fire Bureau
Chapter Seven: Bill Stormont
Chapter Eight: Bill Stormont
Chapter Nine: Bill Stormont

Contents

Preface

In 1975, when I became a career firefighter-EMT in Eugene, Oregon, there were two simple rules that governed our relations with the news media. The first was that if our pictures appeared in the newspaper or on television, we had to buy ice cream for the other members of our companies the following shift. The second was that, at incident scenes, only the chief in charge could talk to the press.

The first rule resulted in a lot of video in which firefighters looked like mobsters leaving the courtroom, their hands covering their faces as they turned away whenever cameras appeared. The second rule forced people who often were too busy, or who didn't want to talk to reporters, to deal with members of the media who showed up at emergencies. Chief officers who were skilled fireground tacticians displayed minimal tact when confronted by people wielding cameras ("Who in hell let you in here!? Get back across the street with the rest of the civilians!"). Those of us who fought the fires would go back to the fire station after a job, change the cotton-jacketed hose, and turn on the news only to be disappointed. The media couldn't get the story right. What eventually became apparent was that the story wasn't right because we weren't talking to them.

This isn't a book for public relations professionals or public information officers in major fire departments. It is a book written for fire service managers in small to midsize departments who find themselves wanting to know more about public and media relations in what has come to be known as the Information Age. It is the type of book I wish I'd had some eighteen years ago when, as a firefighter, I was called into my chief's office and summarily conscripted to serve as the Eugene Fire Department's first public information officer.

America has come to be saturated with news and information. Newspapers, magazines, all-news radio, cable networks, tabloid and reality-based television, and the traditional electronic media provide us with more news and information than ever before in our history. The fire service has long been a source of much of this news: major fires, chemical spills, multiple-patient accidents, and various and sundry other disasters. Many fire chiefs, comfortable in their relationships with the editor of a weekly paper and the news director of the radio station in the county seat, have been overwhelmed by the sudden appearance of satellite trucks and news helicopters when an incident in their jurisdiction makes the national news. If reporters cannot get fast, accurate information from fire officials in such situations, they'll get it from whoever they can—usually an eyewitness who has little understanding of what is really happening.

At the same time that the news media creates risks, it creates opportunities. With its powerful ability to reach mass audiences, it provides fire departments with a tool to cost-effectively provide fire and life safety information to communities and to broadly communicate fire service goals and needs. In major emergencies, the media provides the most efficient way for fire departments to advise citizens about evacuations, road closures, and self-help instructions.

Perhaps most importantly, media relations is a major subset of public relations, an increasingly sophisticated discipline that has found its way into politics, government, and business. The Foundation for American Communications has suggested that "If you don't exist in the media, you don't exist." As an increasing number of institutions and organizations compete for the fleeting glance of the camera, the sound bite, and the favorable editorial, the fire service must also learn how to play this very serious game to maintain its rightful place on the public agenda.

This book is intended to help fire departments master the basics of this game; to help them respond reactively to incidents that place them in the news; and to proactively tell and sell their stories in the court of public opinion. Most importantly, the suggestions made here are for things that departments can do at little cost. In an ideal world, fire departments would be able to hire pollsters or market research firms to do focus groups and detailed statistical inquiry. Interactive computer networks could be set up with local news organizations, as has been done in major metropolitan areas. The reality is that most fire departments have limited resources to devote to public relations and, in the end, the techniques are more important than the technology.

Finally, the lessons taught here have been learned in the hard-knocks school of journalism by a firefighter who, with little formal training, became his department's first PIO, later the acting public information director for his city government, the primary media liaison and spokesman for an administratively merged police and fire department, and, eventually, a chief officer overseeing public affairs for a large suburban fire district. It is my hope that the lessons I've learned are universal and have some benefit for those engaged in the trade that I came to love more than two decades ago.

Chapter One

Public Relations 101

Membership in the fire company has dwindled by ten percent over the past three months, and the chief is worried. Surely there are those in the community who would be willing to serve as volunteers, but how can they best be reached and motivated to join?

A fire marshal looks at the proposed code amendment on his desk. The city council will hold hearings on the measure, an expansion of sprinkler requirements, in two weeks. Local developers are seeking to kill the amendment and have begun a letters-to-the-editor campaign in the town's paper. How can the marshal make his case in the court of public opinion?

An ambulance carrying a critically injured patient to a trauma center experiences a mechanical breakdown. A backup unit responds, but time is lost and the patient is pronounced dead shortly after arrival at the medical facility. The local newspaper and several TV stations want to talk to the EMS manager about the incident. What should the manager say to restore public confidence in the system?

Fire departments, whether they realize it or not, engage in public relations on a daily basis. The kind of reception that callers and visitors to headquarters get, the way fire station tours are handled, the behavior of fire companies on inspection detail, the manner in which news reporters get their questions answered, and the way victims of fires and other emergencies are treated are all examples of public relations that the fire service practices on a daily basis.

Much of what we call public relations is little more than commonsense human relations: treating people the way they want to be treated, responding to their needs, and employing courtesy and tact wherever possible. But while most fire officials have an appreciation of the importance of such hands-on public relations practices, few have any understanding of the range of activities and strategies that PR can offer their departments. Large urban departments often have designated public information officers who generate publicity and handle reporters' inquiries about emergencies and department activities, but most American fire departments leave public communications to chief officers and chance. How best to improve the fire service's ability to do public relations?

To begin at the most basic level, what is public relations? Few professions are subject to as many stereotypes, sometimes negative, as public relations. Mention public relations to some people and the vision that comes to mind is that of a slick, fast-talking man, usually smoking a cigar and wearing a plaid suit. But the days of the old press agent are gone, and the reality of modern PR is much more sophisticated. How do public relations professionals define their craft? Let's consider several definitions, formal and informal.

A public relations professional might describe PR as "communicating specific messages at targeted publics as a means of helping to achieve an organization's goals and objectives." The notion of plural publics is a key one in understanding the fine points of public relations. Often, laypeople assume that PR is a shotgun approach of directing messages to the public at large. Although this is sometimes the case, public relations efforts tend to focus on specific key groups of people. For example, in the recruitment scenario that begins this chapter, the fire chief might want to consider where he could find the kind of people interested in serving as volunteers. By posting flyers in health clubs, he could direct his message to active, physically fit people. By talking to service clubs, he could reach community members who've already demonstrated a commitment to service. Similarly, the fire marshal in the code amendment example might consider several different publics: developers, city council members, local insurance people, and members of the fire department (who often serve as communicators both on and off the job). Each of these groups could be targeted to receive specific messages.

A less formal definition of public relations is that it is the sum total of how an organization relates to the public. What kind of perceptions do community members have of the fire department? What are those perceptions based on? Personal experience? News media coverage?

A reporter once provided a good definition of public relations, contrasting it with journalism. "My job," said the reporter, "is sticking my nose into other people's business. The job of PR is sticking your business under other people's noses."

A good working definition of PR is "the art of doing something well and getting caught at it." If a fire department does its work well, how does the community find out, unless it occasionally gets covered in the news? There are many organizations throughout America still waiting to be discovered, thinking someone will become aware of their good works and cover them in the news. How much easier it is to be discovered if someone on staff has the ability to turn out an occasional news release or otherwise direct a reporter toward a good story.

As a counterpoint to doing things well and getting caught at it, there are some things public relations can't do—like rescue a bad organization. Too often, unenlightened managers believe that PR can put a gloss over serious organizational problems. The reality is that ongoing public relations can enhance the work of a fundamentally sound organization and develop a reservoir of goodwill against the inevitable bad day any organization has from time to time, but the organization has to be fundamentally sound to begin with.

A further definition is in order at this point. The term publicity is often considered synonymous with public relations. Publicity, the process of getting the word out about an issue or upcoming event, is usually the most visible part of the public relations process. But publicity alone does not constitute PR. True public relations is a strategic, ongoing effort.

Just as a fire officer uses a set methodology in responding to and handling an emergency incident, so does the PR professional in handling an issue. The first step, much like the size-up done at a fire, is a fact-finding process. What is the issue? What do people know about it? How is the department positioned in the court of public opinion?

The second step, planning, continues the fireground analogy. What are the department's goals and objectives?

Who are the audiences to be targeted? What messages should they receive? What resources does the department have to prepare and deliver these messages? The planning step is what separates true public relations from simple publicity. The results of the planning phase form a strategy, which should be outlined and documented to direct future efforts.

The third phase, like fire attack, is the active, tactical phase. It involves actual communications, whether news releases, letters to the editor, service club speeches, public displays, inserts in utility billings, meetings with key constituencies, flyers and posters, staged events, or whatever else the communicators come up with.

The fourth and final phase is also one familiar to firefighters in good departments—the postmortem, or critique. What worked? More importantly, what didn't? What lessons were learned for the next time?

When developing a public relations campaign around a particular issue or project, there are several questions that staff should ask:

- What are the department's goals and objectives?

- What do individuals and organizations need to know about each objective for it to be achieved?

- What is the department doing now to communicate with those individuals and organizations?

- What else could be done to reach each individual or group?

- How will it be known when the objectives have been achieved?

- Does the department have the resources to make this happen? If not, where can it find them?

This may seem like a lot, but the process can be as informal as one person making notes on a legal pad or as formal as a brainstorming session at a staff meeting. The bottom line is that true public relations is an organized process with goals, objectives, strategies, and assigned tasks.

So, how does your department begin to make public relations more a part of its operations?

Begin with a hard-nosed audit of what your various publics think of you. Have you gone to the polls with a money measure in the past few years, and did it pass? What kind of relationship do you have with the mayor, city council, commissioners, board of supervisors, and directors? How well do you get along with the local newspapers and broadcasting stations? When your department receives complaints, what are they about?

Evaluate your nonemergency contacts with the community. Are firefighters on inspection detail appropriately dressed, courteous, and properly trained to deal with violations? Does your department provide speakers to community groups that request presentations on fire safety? When groups of schoolchildren tour the fire station, do the firefighters conducting the tours have some key messages to pass along, or are the tours done by rookies because no one really wants to do them? When someone calls headquarters with a question, does he get a friendly reception? Does he get his question answered on the first try? When was the last time your staff had any customer relations training?

Evaluate your emergency contacts with the community. Have your firefighters, EMTs, and paramedics received any kind of basic crisis intervention training to help them console victims? Does your department have a chaplain or community volunteer program that can provide crisis support and advocacy for victims? Has your department put together a brochure to give fire victims information on department operations, postfire cleanup, and replacement of key documents lost in the fire? If someone is left homeless by a fire, do your members know how to connect with social service agencies that can provide shelter and housing? When those social service agencies do fundraising, does your department help with their campaigns or otherwise support them? Are your personnel empowered to "step outside the box" and provide customer service beyond simply handling the immediate emergency? For example, if a motorist were left stranded on a freeway by a car fire, would your personnel let him use a department cell phone to call for help, give him a ride to a safe location, or even provide cab fare?

Public relations objectives can influence such identifiers as apparatus markings and uniforms. The current trend of dressing American firefighters in uniform T-shirts with department names boldly printed across the back developed, in part, because of television. As more and more departments became EMS providers, they found themselves appearing on the evening news without turnouts. Since fire uniforms are often similar to those of police, security, and private-sector ambulance providers, what better way to announce who's providing the service than to have the department identity appear like a small billboard on the back of every member working a call?

Apparatus markings are another way of ensuring that service recipients know what agency is serving them. Is your department's identity boldly marked on each piece of apparatus? Some departments, recognizing that most citizens see all apparatus merely as fire trucks, go so far as to clearly mark each rig by type (i.e., ladder, engine, rescue, or medic), thus clearly identifying to nonfire folks what sort of rig they're seeing.

While a public relations purist would argue correctly that PR should be a part of all department policies, several key areas should be addressed specifically in department policy. These include news media relations, fire station tours, and the handling of complaints.

Finally, decisions must be made about staffing your department's public relations effort. No matter how the work is staffed, two things are true. First, all department members must understand that they collectively carry most of the responsibility for how their organization is regarded in the community. Second, no matter how well intentioned, any effort is doomed to fail unless it has strong support from the chief and the management team. There are as many ways to staff the public relations function as there are fire departments. Some departments have standing PR committees, others use volunteers with strong communication skills, and some assign responsibility to a particular officer or firefighter along with his other duties. I strongly recommend that your department appoint one person, full or part time, paid or volunteer, to serve as public information officer (PIO). At minimum, this person should serve as the primary liaison with the news media to maintain a positive ongoing relationship and provide the community with a steady stream of information about the department. The chief, the management team, or a department public relations committee can provide suggestions and input, but it's best to have one person assigned to public relations, seeing that the day-to-day work gets done and answering routine media inquiries about fires, rescues, and other emergencies. In a later chapter, we'll look at how to select and prepare the person who will be PIO, and the balance of this book will provide the PIO with the tools to do his job.

Chapter Two

Who Are These Reporters and What Do They Want, Anyway?

The fire marshal picked up the report and looked across his desk at the PIO, asking, "How much coverage are the annual fire statistics going to get?"

"Well," she replied, "the paper will probably print just about everything in the report. TV will give us thirty seconds and use file footage from the mill fire last summer. We need to decide on one or two key points we want to make to the radio stations."

In many ways the media are all alike, and in many ways they're different. The successful PIO must be a student of the media and learn as much as possible about who journalists are and how they work.

The media weren't always "the media." At one time in our history, they were "the press." With the growth of television and radio broadcasting, however, an increasing number of news organizations no longer owned presses, and so the terminology changed. As part of the same change, press conferences were replaced by news conferences, and news releases replaced press releases.

It's important to have an understanding of the role of the media in American history to grasp the political and constitutional framework under which they operate. It's not my intent to provide you with a detailed history of the media in America or an extensive analysis of media theory and criticism. If you are interested, any number of texts can provide additional background. Suffice it to say, however, that anyone who works regularly with the media needs to understand a few history-driven principles.

News organizations are unique institutions in American society. On one hand, they are privately owned, for-profit businesses. On the other hand, they are accorded constitutional protections and powers like those of public institutions.

The framers of the American constitution, cognizant of human frailties in governance, saw the benefit of having a press that, free of government restriction, could report honestly and openly on public affairs without fear of official censorship. The early leaders of the republic saw the press as playing a key role in providing citizens with the information they needed to govern themselves and so accorded the press significant legal protection and standing.

To quote James Madison, "Knowledge will forever govern ignorance. And a people who mean to be their own governors must arm themselves with the power knowledge gives. A popular government without proper information or the means of acquiring it is but a prologue to a farce, or a tragedy, or perhaps both."

While recognizing the occasional need to delay disclosure of government information, Patrick Henry indicated strong support of a free press when he said, "I am not an advocate for divulging indiscriminately all the operations of government, though the practices of our ancestors, in some degree, justify it. Such transactions as relate to military operations or affairs of great consequence, the immediate promulgation of which might defeat the interests of the community, I would not wish to be published, till the end which required their secrecy should have been effected. But to cover with the veil of secrecy the common routine of business is an abomination in the eyes of every intelligent man."

These philosophical and constitutional underpinnings have continued to this day, although it would be interesting to see our founding fathers' reaction to tabloid television and newspapers, pornography, and some of the more outrageous programming found on public access TV.

A key theme in the theory of the American news media is that it serves as a watchdog over government, covering public meetings, events, and issues to report to the populace on how well the public sector is doing its job. This watchdog role sets up a certain amount of tension, sometimes conflict, between public officials and the journalists who cover their activities.

The newspapers of America's first century were often frankly biased and aligned with political parties, but as the nation moved into the latter part of the 1800s, newspapers gradually became more objective and neutral in their reporting. Part of this trend came from the establishment of wire services. In 1848, the Associated Press (AP) was created to produce news for, and share news from, newspapers subscribing to its services. Because the AP material was produced to be used by papers of all political stripes, the copy was very neutral and objective.

This trend toward objectivity in news reporting spread throughout most mainstream daily newspapers. Editorial opinion and political slant remained on the editorial pages, where newspapers endorsed candidates for public office and ballot measures, attempting to sway public opinion, but the news in the papers became increasingly fact-based and objective. Newspapers became less the voices of political parties and more competing businesses trying to capture the largest share of readers in their markets.

While newspapers separated news from opinion in the 20th century, they were joined by new forms of mass media: first radio, then television. Radio revolutionized news reporting with its sheer immediacy, and television brought moving images of news events into our lives.

In recent history, the media have undergone even more dramatic and sometimes controversial changes. The dominance of the big three television networks (ABC, CBS, and NBC) has been challenged by the lightning growth of cable television, which offers a mind-numbing smorgasbord of choices for viewers, including channels in such specialized niches as sports, health, history, and music videos. Television news is provided on a 24-hour basis by CNN, a format that is now being copied locally in major urban areas. All-news coverage is also provided by cable outlets specializing in financial reporting and public affairs, such as C-SPAN. Local broadcast news organizations have embraced technology that enables them to go live to the scenes of major events, creating logistical demands on public safety officials who work with them.

Print journalism has undergone revolutionary change as well. The magazine industry, said to be on its deathbed in the 1970s, has expanded dramatically as new publications have mined the interests of increasingly narrow and specialized readerships. Satellite technology has enabled the regional printing and distribution of *USA Today*, the first truly national daily newspaper, and other major metropolitan dailies have used the same technology to produce regional or national editions, as *The New York Times* has done. The alternative press, ranging from underground newspapers to slick journals of opinion, has continued to provide investigative reporting and alternative opinions and viewpoints to those found in the mainstream press.

Radio has enjoyed a renaissance, too, with different stations targeting listeners in specific age and income brackets. News radio and talk radio have emerged as major forces in American politics in the '90s, with nationally syndicated hosts on both the political left and right drawing huge followings. Nearly every large American city now has at least one all-news radio station.

In all forms of journalism, a new trend has been toward "info-tainment," and it has become increasingly difficult to draw the line between straight news and other forms of journalism. Consider the rise of daytime television talk shows on which people share sometimes peculiar and intimate details of their lives; the growing number of magazine-style shows that focus on celebrities; and the reality-based shows that bring us the activities of emergency responders in different parts of the country. All of these are forms of media, all of them have different interests and needs, and many of them have contacted local agency PIOs for help on different stories.

For now, let's get back to the basics and focus on daily

news in our local communities—the kind of occurrences that PIOs deal with on a daily basis. To begin with, what is news?

When we talk about news, we're really talking about news value, the definition of which varies from community to community and day to day. There are weekly papers in small American communities that report the visits of out-of-town relatives, elections of officers in garden clubs, and every police and fire call. The news threshold in big cities is set higher, and news is limited to major events and political occurrences—although feature reporting often focuses on the noteworthy activities of small numbers of people and human interest stories. What this means for the fire service is that every fire is considered news in some places, while only major fires get covered in the larger cities.

But news value isn't shaped just by community size alone. News value varies daily, depending on what else is going on in the community. An old axiom among PR people is "Never mess up on a slow news day." Consider, for example, a local television station that does a one-hour newscast every night. If, in a single day, the community experiences the resignation of a politician, the closure of a factory, a two-alarm fire, and a homicide, then the TV station will have a pretty full newscast. If, on another day, few major events happen within the community, the TV station will still have an hour's worth of airtime to fill, but lesser news will fill it. One day, your event gets covered; the next day, nobody shows up. Remember, your news is always in competition with other stories of the day.

Generally speaking, what makes news? There is no hard-and-fast definition, but some of the factors that editors and producers consider when determining the merits of a story include:

How many people does it affect? Does the story have general or broad community interest? The activities of a city council should be of concern to all residents of a community; those of a CB radio club won't. To be included in the news, a story has to captivate, cater, or appeal to at least a substantial minority of the audience.

Is the story controversial? Since the first time we heard in grade school that there was going to be a fight on the playground, most of us have been interested in conflict. What's the issue? Who's pro? Who's con? What are the merits of their arguments? These are often easy stories to do—the reporter offers an introduction neutrally summarizing the issue, follows with emotional quotes from proponents and opponents, and closes by saying that only time will tell which side will win.

Does the story deal with real change? When fire departments expanded their roles to include emergency medical services in the '70s and '80s, the change represented a major institutional adjustment and so got coverage in most of the communities where it occurred. Similarly, new methods of service delivery tend to attract news coverage.

Is the story about something new? Remember when CPR was first taught to citizens, and every time someone saved a life by using the technique it got covered in the local news? That doesn't happen so much these days, even though people are still saving lives with CPR. What's happened is that it's become so common that it isn't necessarily considered newsworthy anymore.

Is the story timely? A dramatic rescue isn't as newsworthy two days after it took place, which is why fire departments must develop ways to publicize such events when they actually happen.

Does the story involve public funds? Since all citizens pay taxes, they all have an interest in how their money is being spent. If you're engaging in a major project to build new fire stations, purchase apparatus, or start a new program, it will be news.

Does the story involve someone of prominence? A visit by the chief's brother isn't news, but a visit by the governor probably will be.

Is the story about bad news? The fact that 50,000 people uneventfully commuted into the city this morning isn't news, but that six of them died in a fiery four-car accident is. For better or worse, news is about the aberrations, the large and small tragedies that make up daily life. Most of the time, when public safety agencies find themselves in the news, it's because something bad has happened. It's often argued that the news shouldn't have so much crisis and crime in it, but ratings and circulation figures indicate that people have a lot of interest in such things. Bad things happen, news organizations have a right to cover them, and public safety officials have a responsibility to answer media questions about them. In many cases, it's critical to tell people what to do or not to do in certain situations, and tragedies often provide the impetus for public education or code changes.

What are the basic elements of any news story? Who, what, when, where, how, and why. Don't overlook "how much" if the story is about the expenditure of public funds. These questions cover the essential elements of any news story and will normally be included in the first sentence or paragraph of a news item. You should also bear them in mind anytime you're writing a news release or preparing for an interview. To illustrate: "A Springfield man (who) was rescued by firefighters (what) early Friday morning (when) from his burning Main Street apartment (where). John Stone, 45, was rescued through an upstairs window (how) after a carelessly discarded cigarette ignited a fire in a front room couch (why)."

These are the basics of news, but they are attended to differently by each of the three major media: print, television, and radio. Let's look at the differences between them.

Newspapers provide permanent records of events and offer the highest level of detail of any medium. If you were to take the script, the actual words spoken by the anchors, reporters, and news sources on a thirty-minute network television newscast, and set those words in newspaper-style type, they would barely cover a single page of a newspaper. Newspapers want details—not just the full names but also the ages and addresses of the people they cover.

The permanency of type is another quality that affects how people view newspapers. Television and radio newscasts come and go in a matter of seconds. If you weren't in the room when a story aired, you don't know about it. But newspaper stories last forever. They get photocopied, cut out and put on bulletin boards, mailed to relatives, and placed in scrapbooks. Negative stories can stare up at us from desks and coffee tables for days at a time. The library keeps the newspaper on microfilm or in a database, where it can be studied by researchers, scholars, biographers, and historians. Such permanence is the essence of print.

Because of this permanence, the need to correct factual errors in newspaper stories can also be critical. A reporter assigned to write something new about a long-running issue will commonly go back through the newspaper's "morgue," or library, and look at what has previously been written on the topic. If the record contains errors, many of those errors may well be repeated in new stories.

Newspapers lack the immediacy of delivery that characterizes television and radio. Typically they have a single daily or, in smaller communities, weekly deadline. Newspapers compensate for this by providing more depth and detail in their coverage of events and often require a higher degree of specialization among their reporters. Larger papers, for example, often assign a single reporter to cover just police and fire activity. Thus, the reporter becomes better known to the people he covers and develops more expertise about how public safety agencies do their work. Often, these police beat reporters, as they are commonly known, will go to police and fire stations on a daily basis and look over whatever logs, reports, and other documents are available for public inspection.

Newspaper interviews are usually the least intimidating, done as one-on-one conversations, with the reporter taking notes or perhaps using a tape recorder. Sometimes the interviews are conducted over the phone, in which case spokespersons should always assume they are being taped. Many states have laws that allow the unannounced taping of phone conversations, just as long as one party to the conversation knows it's being recorded.

The key players at most newspapers include the managing editor, who is the head newsroom administrator; city editors, who make assignments and oversee the coverage of local news; reporters, with whom news sources have the most contact; and photographers, who frequently turn up at scheduled events and emergency incidents. Most newspapers maintain a separation between the people who cover the news and those who prepare the editorial pages. If your department expects to seek editorial endorsements or opinions, it's a good idea to get to know who the editorial page staffers are.

Radio is the most mobile medium and requires the least of its audience. It is passive and can be consumed while driving, jogging, or working. Radio stations specifically target their audiences by format or the type of music they play, and they keep detailed listener demographic data to woo potential advertisers. At one time, radio was believed to be a dying medium, but such has not proved to be the case. Talk radio, in particular, has been a rapidly growing phenomenon, offering growing opportunities to publicize issues and activities.

One of radio's biggest values to fire departments and other emergency services is its ability to communicate instantly when something happens. If an emergency results in the closure of a major street or requires that an area be evacuated, the first calls that a savvy PIO makes will be to the local radio stations. The newspaper won't publish its account until tomorrow, TV won't get the story on until five o'clock (unless it's something really major), but radio can get the word out right now to people living, working, and driving in the area.

This is especially true when you may be trying to get important information out to people in the wake of a calamity such as an earthquake, when reminders about what to do and what not to do can save lives. The Emergency Broadcast System (EBS) was established during the Cold War as a way to communicate with the populace during a national crisis. Although the familiar tests are still heard across the radio dial, the system has fallen into a state of disrepair in some communities, while others have broadened its use for all different kinds of emergencies. It's important to find out what the state of EBS is in your community and know how to access it.

The frequency of radio news varies from station to station, and some radio stations don't do news at all. The stations that do news typically do so on an hourly or half-hourly basis, and often much of their news is presented during morning and evening drive times—the commuting hours, when stations typically find their largest audiences.

Most radio news staffs are very small, sometimes no more than a single person; sometimes merely a disk jockey who rips and reads material from a wire service machine. For this reason, most of their interviews are taped over the phone. In larger communities, radio reporters occasionally show up at emergency incidents, news conferences, and other events, but it's best to plan to feed them what they need over the phone.

Radio news is abbreviated, and radio stories are often no more than a couple of sentences in length. But those who write news for radio have mastered the art of jamming a lot of information into those few sentences. ("Bob Smith took his campaign for governor into the rural downstate Tuesday. The two-term state senator from the Tri-Cities told farmers he would continue funding for the state's extension service.")

As a news source, you must learn to meet the needs of

radio. Most importantly, you must learn the art of summarizing any story or the key points you want to make in about fourteen seconds or less.

Another service that radio and television stations can provide you is the airing of public service announcements (PSAs). A PSA is nothing more than a short spot of ten, thirty, or sixty seconds in length that gives people information about government or nonprofit programs and is aired without charge. The best PSAs are those that promote positive behaviors in such areas as fire and traffic safety, to give two examples. There are no requirements that stations use any particular PSA, but the more relevant they are to community issues and the better they are written, the more likely they will be aired. Some radio stations want PSAs on tape and ready for use; others want the spots in written format and prefer to record them with voices from their own staff. If you decide to do a PSA, check with your local station as to how it wants spots submitted. In some cases, if you can convince the staff at the station that the PSAs will address a serious problem in the community, they will help you produce them.

The key personnel to know at any radio station include the station manager, who oversees all the station's operations; the news director, who supervises news coverage; reporters; and the public service director, who handles PSAs and sometimes the booking of guests for public affairs talk shows.

Television is probably the single most powerful medium in the news industry. A growing number of Americans report that they get most of their news from television. Television combines graphics, the spoken word, and moving images into a powerful elixir. Although television is acknowledged as having a great impact on public affairs, the medium's critics point out that TV is often long on visuals and short on information, making it a headline service that oversimplifies stories to fit its time constraints.

One of the reasons that the fire service, like law enforcement, finds itself on television so often is that emergencies are so visually compelling: burning objects, explosions, flashing lights, drama, and action. Incident stories can also be told in less than ninety seconds. Television doesn't do as well when it tries to explain the complexities of municipal budgets or fire code amendments.

Television brings with it a plethora of side issues when it shows up at incident scenes. Microwave and satellite trucks take up space and need to be positioned for line-of-sight access to transmitters. People who wouldn't notice the presence of a newspaper reporter with a notepad change their behavior entirely at the sight of vans with brightly colored station logos and their load of cameras, lights, and microphones. Chief officers who experience no qualms whatsoever when talking with newspaper reporters may refuse to go on camera, convinced they'll come across as inarticulate—or worse. Eleven o'clock approaches, and two or three TV crews all ask for a live interview at the top of their newscast. How does the camera-shy chief react then?

At the same time, television communicates with an immediacy that, in a major emergency, can rival that of radio. It has the same capacity to pass along important information in a timely manner and it lets people see, as well as hear, just what it is that's going on.

Because television is so driven by visuals, it's important to plan accordingly when soliciting TV coverage of a story or issue. It's a show-and-tell medium. The location where you hold staged events, the props you use to tell your story—all should be planned to make for interesting television. If, for example, you're doing a story on smoke detector maintenance, be prepared to take TV crews into a home and demonstrate visually the key points of detector placement and maintenance. Members of my family have been used shamelessly as actors in stories on home fire escape planning and water safety. If you've just acquired a state-of-the-art aerial ladder truck, be prepared to take a photographer to the ladder's tip and show what a flowing master stream looks like.

Television, like radio, can also help with PSAs,

although TV PSAs are more expensive to produce. In some cases, you may be able to get preproduced spots from other sources and convince your local stations to add your tag, or department ID, and run them. As with radio, some stations may be willing to devote resources to helping you produce the spots, assuming you can convince them of the community interest and need.

Another opportunity is the multipart series that runs on the local news on consecutive nights. It's not uncommon, for example, during Fire Prevention Week for a local TV station to devote a few minutes each night to a series of stories on different aspects of fire safety. Similar series have been done on fire department EMS programs, 911, and other public safety topics.

The key players to get to know at a television station include the news director, who manages all news operations; producers, who are responsible for overseeing the assembly of newscasts or other programs; assignment editors, who serve as newsroom dispatchers and coordinate the response of crews to scheduled events and breaking incidents, as well as set up interviews with news sources; reporters, who typically cover several stories in the course of their working days; and camerapersons who, given the visual nature of their medium, work in partnership with the reporters to whom they're assigned.

A few words are in order about reporters. In many ways, reporters are no different from firefighters. Some are bright, some are dull. Some are kind, some will challenge your patience. Some are funny and warm, some are serious and cold. What they all have in common is that they are professionals with a job to do and bills to pay. In this day and age, most of them have a college degree and a particular set of skills. What they are good at is taking a body of information, boiling it down, and making it interesting to read, hear, or watch for the people who consume whatever product their medium creates.

Like many other groups of people, reporters have been stereotyped in often negative and inaccurate ways. In nearly two decades of working with journalists, I've found the lion's share of my relationships with news people to be professional and positive; a number have turned into warm, long-term friendships.

Always remember that, while reporters may be bright, competent people, they may have little knowledge about what you do and how you do it. Unless you know a reporter well enough to respect his overall knowledge, be patient and take the time to explain the elementary concepts, such as why firefighters chop holes in the roofs of burning buildings and how those tanks on their backs carry compressed air, not oxygen. Your nonjudgmental patience and willingness to explain will pay dividends in long-term relationships and accurate reporting.

A basic tool that you will need is a media list, a directory of local news organizations with specific information about each one. At minimum, that information should include each newsroom's mailing address, phone and fax numbers, after-hours numbers for key personnel, the names and numbers of public service directors at TV and radio stations, information on newscast times and deadlines, and such special PR opportunities as community calendars and public affairs programs. In short, the media list should be a compendium of the basic information you'll need to know about every news organization in your community.

Chapter Three

Communicating With the Communicators

The chief moved about his office with excitement and enthusiasm. It wasn't often that the PIO had seen him like this. He pulled some papers from a routing box on his desk and said, "Here's the deal. The city manager has given me the go-ahead to pursue construction of two new fire stations, provided the city council will give us the funding. I'm making a presentation to the council Monday night, but I want to get as much coverage as I can for the proposal. How do you suggest we proceed?"

While much of the work of PIOs is reactive, providing information about emergency incidents, a certain amount of their work is proactive, initiating coverage of department activities, plans, and projects. When an incident occurs, the media typically come to the department with their inquiries, but how should a department approach the media?

The first step is in deciding whether or not you have a legitimate news story. Ironically, the PIO who pleases the boss by putting out a steady stream of releases may be developing a newsroom reputation for crying wolf if the tips aren't really news. In the previous chapter, we discussed news value and how it varies from community to community and day to day. Part of a PIO's value to the department is his ability to exercise judgment. The training chief may believe that the new helmet markings are the greatest invention since the fog nozzle, but the average reader of the morning paper probably couldn't care less. (On the other hand, if the helmet markings are truly innovative, an enterprising PIO might write up something for one of the fire trade magazines.) It is important for a PIO to be a regular consumer of local news and be up to speed on what kinds of stories are getting editorial attention and focus. Don't overlook the possibility of getting local coverage on national stories. Many departments around the country used the 1980 back-to-back fires at the MGM Grand and Las Vegas Hilton as an opportunity to generate stories on high-rise fire safety or to push local code amendments. The tragic Happy Land Social Club fire in New York City's South Bronx provided fire officials across the country with an opportunity to explain fire codes in places of public assembly and provide safety tips for readers, viewers, and listeners who frequent such places. The Oakland Hills Fire and the rural/urban interface firestorms around Los Angeles resulted in calls to fire departments throughout the West from news organizations wanting to do "Could it happen here?" stories.

Factors to consider in determining the newsworthiness of a story include the number of people it will affect, the potential for controversy, the amount of public money that will be spent, whether it concerns an issue already on the public agenda, whether it involves a new approach to an old problem, basic human interest, and (in concession to television) whether it has compelling visuals. The fire station construction example that begins this chapter meets most of these criteria and is certainly one for which the department should initiate coverage.

The first decision that must be made is who is going to speak for the department. Ideally, the chief would be the best source, and many chiefs want to do all the public speaking for their departments. The problem arises when the chief is too busy to be always available or, frankly, when the chief isn't comfortable talking to reporters. In some cases, such as the fire station construction example, it *ought* to be the chief. Reporters expect the boss to be accessible when major news, good or bad, happens. The role of the PIO in such cases is to be a facilitator for the chief: to help anticipate the questions that will come up, to coach the boss on making key points so they won't be overlooked, to give reporters adequate background information so they won't waste the chief's time, and to arrange and schedule interviews.

The chief doesn't always have to be the spokesperson. One of the reasons large departments have PIOs is that the volume of media requests for information and interviews mandates a full-time position for someone to field them, especially if the requests are fairly routine in nature. In many cases, the PIO can arrange for other staff to do the interviews. A story on Christmas fire safety, for example, might best be handled by a member of the fire prevention bureau. Part of the PIO's job is to connect reporters with the best sources of information for interviews and to prepare those sources so that the results are as positive and productive as possible.

With the spokesperson identified, how do you get the word out? The most common method is to prepare and send out a news release. A news release is little more than a short memo, written in journalistic style like an article, that tells newsrooms about an upcoming event or an issue, providing them with the information they need to follow up should they decide to cover the story. Formats for news releases vary but always include the following elements:

They're typewritten on letterhead stationary.

They have a heading that reads *News release* and includes the name and phone number of a contact person so that newsrooms wishing to follow up on the story can easily get ahold of a knowledgeable spokesperson.

They have the date of issue in the heading.

They either say *For immediate release* or have a release date and time as part of the heading. *For immediate release* indicates that newsrooms may report the information in the release as soon as they receive it. A release date and time is a request that information not be publicized until that time. For example, if a disaster drill is planned but details are being kept secret from department personnel, a news release may be prepared and sent to newsrooms in advance of the drill, with a release date and time asking that the information not be reported until the morning of the exercise.

They have a brief title, or headline, that explains what they're about; for example, "Fire Department Plans Public CPR Classes."

The body of the release is typed double-spaced, with wide

margins, allowing editors to mark them up and make notes before passing them on to reporters assigned to the story.

Keep the release to one page, if possible, and certainly to no more than two or three. Remember, you only need to provide the basic who, what, when, where, how, and why in a news release, not the whole story as it will be published or aired.

If the release is longer than one page, type *More* at the page bottoms to indicate there's more material to follow. At the top of successive pages, type a brief heading and page number, such as CPR-2.

At the end of the release, type either 30 or ###. These symbols have long been used in journalism to signify the end of a story. Legend has it that they were first used when reporters sent their stories over telegraph wires.

Most importantly, learn to write in journalistic style. Summarize your story in the first sentence, known as the lead, stressing the most important points. The average newsroom receives hundreds of news releases a week, and the editors who review them skim and read quickly. If you can get your point across and capture their interest in the first sentence or two, you'll have won half the battle.

The journalistic style of writing is called "inverted pyramid" writing, in which all of the main points are up front in a story, with lesser details bringing up the rear. The style found its place in journalism so that, if a piece had to be cut to make it fit, a busy editor could simply cut successive paragraphs from the bottom without harming the main thrust of the story.

In the words of *Dragnet's* Sergeant Friday, write "just the facts." Write tightly, and leave out hyperbole and editorial comment. Let the facts tell the story. Check the facts, numbers, and spellings closely. Be specific. Don't write "many" or "some" if you know the exact number. Avoid fire/EMS jargon and stick to plain language that will be comprehensible to non-fire service readers. Use facts to fill space, and when you run out of facts, stop writing.

There are several ways to learn how to write in this style.

One way is simply to read news stories with an eye toward their format as well as their content. Another way to learn more about newswriting is to read journalism textbooks, which can usually be found in libraries or even used bookstores.

Basic references useful in writing news releases include a dictionary, a thesaurus, and a style book—a text that contains rules for punctuation, capitalization, abbreviation, usage, and a host of other useful entries. Many different news organizations have their own style manuals so their stories will be consistent in style and grammar. One of the easiest to find and most useful of journalistic style manuals is *The Associated Press Stylebook and Libel Manual*.

Two actual news releases are reproduced on the following pages as examples of format and writing style.

Once the news release has been written, then what? Have it reviewed by other members of the department to make sure it's accurate and understandable. If it's major news, or if policy requires, make sure the chief has a chance to review it before it goes out. When the release is ready, deliver it to your targeted newsrooms using the media list discussed at the end of Chapter Two.

What's the best way to deliver your news release? It depends on the preferences of your local newsrooms and how quickly you want to get the word out. If it's not really major news and you can easily meet the lead times asked for by newsrooms on your media list, first-class mail will work fine. If you have the time, if it's a big story, or if you're in a small community, you can hand deliver the releases. If your department has a fax machine, you can send the releases immediately. Some departments have fax machines in which all of the local newsroom fax numbers can be preprogrammed as a set. By hitting a single button, the machine will automatically transmit the release to all of the local newsrooms.

A general rule with news releases is to choose a distribution method that allows them to reach all newsrooms at about the same time. This is especially important if you're

TUALATIN VALLEY FIRE AND RESCUE
NEWS RELEASE
February 10, 1997
Contact: Tim Birr, 503/555-5555
FOR IMMEDIATE RELEASE

Fire Damages Beaverton Auto Dealership

A fire in a showroom ceiling did an estimated $50,000 dam-
age to the Herzog-Meier Autocenter in Beaverton Monday
evening. Tualatin Valley Fire and Rescue crews were dispatched to the dealership after employees working in the
building called 911 at 7:26 p.m. to report light smoke
appearing across the showroom ceiling. When crews arrived,
they entered the building, located at 139th Way and Tualatin
Valley Highway, and attempted to determine the source of the
smoke. The smoke continued to build in density, and employees were evacuated from the building as firefighters discovered fire in the space between the showroom ceiling and
the roof of the dealership.

Given the size of the building and concerns about rapid
fire spread, a second alarm was called, bringing additional crews and equipment to the scene at 7:37 p.m.
Firefighters discovered that the fire was burning in the
insulation above the showroom ceiling. Crews quickly covered
the vehicles parked in the showroom with protective tarps
and began pulling down ceiling tiles and smoldering insulation. A third alarm was called at 7:52 p.m. to bring additional crews to the scene to assist with the laborious job
of pulling down the showroom ceiling. The fire was declared
under control at 8:13 p.m.

The fire was confined to the insulation that rested above
the ceiling and appeared to have spread along paper backing
on the insulation material. Fire investigators were still
examining electrical wiring in the area above the ceiling
and a fireplace flue in the showroom but had not yet determined a cause for the fire late Monday night.

Twenty units and an estimated 60 firefighters responded to
the incident. There were no injuries to building occupants.
One firefighter was treated at the scene after some debris
got into his eye during the ceiling-pulling operation.

###

TUALATIN VALLEY FIRE AND RESCUE
NEWS RELEASE
January 22, 1997
Contact: Karen Eubanks, 503/555-5555
FOR IMMEDIATE RELEASE

DISTRICT ISSUES REMINDER ABOUT DANGER OF WARM ASHES

   Since the arrival of cooler weather, Tualatin Valley
Fire and Rescue has responded to a multitude of fires
involving warm ashes disposed of improperly. The biggest
incident occurred on December 27 and involved a $1.4-mil-
lion home that sustained more than $100,000 in damage due
to ashes from a fireplace placed in a garbage sack next to
the home. In two other incidents, warm ashes were repon-
sible for destroying outside shop buildings, valued at
$7,000 and $17,000.
   The district reminds residents that, although ashes may
appear or feel cool on the surface, they can remain warm
enough inside to start a fire two to three days after they
have been removed from a woodstove or fireplace. If residents cannot wait several days before cleaning out a stove
or fireplace, they should place the ashes in a closed metal
container outside the home, well away from combustibles.
Residents can also wet down ashes with a garden hose to
ensure that they are completely extinguished.

###

dealing with a major story that's marked for immediate release. The news business is very competitive, and if anyone thinks you're giving the competition an edge by selectively releasing information, it can hurt your relationship.

If your news is of major significance, don't forget to brief and share copies with any elected officials who oversee your agency. Few things upset mayors, council members, and commissioners more than being called for comment or reaction to matters their own staffs haven't briefed them on.

Additionally, if your department serves an area with a large non-English-speaking population, don't overlook communicating with them. As I write this, there is ongoing debate in different parts of the country as to whether English should be the official language of the United States and to what extent public agencies should develop multilingual capacities. My viewpoint is pragmatic. Such populations exist in growing numbers around the country, and we create liabilities for ourselves when we don't communicate with them. Conversely, we create positive relationships when we make the effort to ensure that they receive the same safety information as the community at large.

Consider a major fire in a building or neighborhood populated with non-English-speaking people. If your department doesn't have the ability to communicate with the victims, survivors, witnesses, and neighbors, you have no way of knowing what kind of rumors, gossip, and misinformation might be spreading throughout the community about the incident and your agency's efforts. If the news media's ahead of your department in terms of multicultural awareness, you may be unpleasantly surprised when a bilingual journalist hears and reports on matters of which you had no knowledge. It doesn't take much to anticipate the question, "This city has a large ethnic population. What steps has your department taken to educate this community on basic fire safety?"

Consider a disaster such as an earthquake or a major storm. When self-help information is critical, all of the "what to do" and "what not to do" advice you put out needs to reach the entire community. A lot can be done to prepare such information in advance, and you're better off if you have the tools in hand to reach your minority populations before disaster strikes.

Affirmative action, quotas, and politics aside, these issues offer compelling reasons for fire departments to ensure that their memberships reflect the communities they serve. It's easier to communicate with minority populations when you have people in your organization who speak the language and understand the culture.

Even if you lack language skills, there are several steps you can take. Most areas that have large non-English-speaking populations have minority media serving those markets. Such sources may take the form of weekly or monthly newspapers, newsletters, cable television, or radio. In some cases, broadcast media that do most of their programming in English may dedicate part of their airtime to foreign-language material. It's important to include these news organizations on your media lists and ensure that they hear from you on a regular basis. While it's probably not critical that they get your news release on a new ladder truck, it is important that they receive regular safety advice and information on incidents that affect the populations they serve. If there's no one on staff who can help with translating it, you're likely to find that the editors and producers of these publications and broadcasts can help you. Part of their work consists of taking information from the English-speaking community and preparing it for their particular audience. Another possibility, if you're developing information for distribution to a minority community, is to obtain the services of a college intern or to enlist the assistance of leaders in that community.

It is important to recognize the existence of the community and make the effort to communicate with its citizens. As large as some of these communities are, it's amazing how often fire departments fail to communicate with them. Making the effort pays dividends in community support and public safety.

What about emergency incidents? What if there isn't time to write a standard news release from scratch? The easiest way to handle such a situation, and one used by many departments around the country, is to prepare standard news release forms for the more common types of incidents your department handles. The blanks on the forms, which are basically abbreviated incident reports, can be filled in at emergency scenes, referred to as notes for media interviews, and, provided your handwriting is legible, faxed to newsrooms that didn't send reporters to the incident.

You can prepare as many types of incident release forms as you want. Most departments that have developed forms have at least one for fires and perhaps others for situations such as hazardous materials incidents. If you've designed them to suit your needs, you'll find that they save time at the scene and enable you to gather information without having to stop and think about what it is you have to gather. Examples of fire and haz mat forms are shown on the following pages to give you some idea of what key elements should be included.

You can also use the forms as scripts to record brief reports on a telephone answering machine, the number to which has been previously provided to local newsrooms. An answering machine can be used to disseminate news releases, traffic advisories, and seasonal safety tips. The best machines for this purpose have a long outgoing message feature, enabling you to record several minutes' of material. It's best if the machine is designed so that the recording can be changed remotely, such as by cellular telephone from an emergency scene. Local newsrooms can routinely check such machines for information, eliminating the need for a lot of repetitive follow-up phone calls.

Another type of news release used by some emergency service agencies is the prewritten, fill-in-the-blank type used to give the public lifesaving instructions in the event of a major disaster. Let's say, for example, that your community is hit by an earthquake. The phone rings immedi-

TUALATIN VALLEY FIRE AND RESCUE
Karen Eubanks, Public Information Officer
503/555-5555

EMERGENCY INCIDENT NEWS RELEASE

1. Type of Incident: _____

2. Address of Incident: _____

3. Date: _____ Alarm Time: _____ Recall: _____

4. Description of Occupancy: _____

5. Names (Firm, Owner, Tenancy): _____

6. Visual Description on Arrival: _____

7. Probable Cause and Point of Origin: _____

8. Estimated Loss: Building: ____ Contents: ___ Equipment: ____

9. Assignment: Engine Companies _____ Rescue:_____
               Truck Companies _____ Other: _____
               Total Manpower:_____

10. Mutual Aid: _____

11. Injuries, Rescues (Name, Age, Address), and Items of
    Human Interest: _____
    _____
    _____
    _____
    _____
    _____
    _____

EUGENE DEPARTMENT OF PUBLIC SAFETY
HAZ MAT INCIDENT NEWS RELEASE

Date: _____

Location: _____

Times: Dispatch:_____Arrival: _____Control: _____

What Happened: _____

_____

Name of Involved Product: _____

What the Product Is Used for: _____

Characteristics and Hazards: _____

Current Situation as Verified by Facts: _____

_____

Public Impacts (Evacuations, Street Closures, etc.):

_____

_____

Number Killed/Injured/Affected: _____

Agencies on Scene: _____

_____

Incident Commander: _____

Additional Information/Notes:

_____

_____

_____

ately—a local radio station is seeking information. What can you say? Well, if you've prepared in advance, you could have a prewritten news release that reads something like this:

"This is _____ at the _____ Fire Department. At _____ a.m./p.m., an earthquake was felt in the _____ area. At this time, we have no confirmed reports of injuries or damage. Area residents are reminded not to call 911 unless they have an actual emergency to report, and they are also asked not to use phones except in emergencies. In the event of aftershocks, residents are urged to seek shelter under a table or desk, cover their head, and hold on. If your house has been damaged and you smell gas, shut off the main valve. Switch off electrical power if you suspect damage to electrical wiring. If you are outdoors, stay away from buildings, trees, and overhead utility lines. Listen to local radio stations for more information, which will be provided as it becomes available."

These prewritten releases can be recorded in series, with follow-up releases advising citizens of quake magnitude, road and bridge closures, the status of schoolchildren, and other essential information. Similar releases can be prepared in advance for whatever emergencies your community faces, including hazardous materials incidents, severe weather, wildfires, and floods.

Emergencies aside, let's get back to the kind of proactive work discussed at the beginning of this chapter. If news releases are one staple of the public relations trade, news conferences are the other. A news conference is an event set up to make an announcement, to introduce to the media someone with a story or expertise on a given topic, or to kick off a campaign or project. As with a news release, it's important to make sure the story is big enough to warrant staging such an event.

One problem with news conferences is that they put you

in direct competition with any other news story going on at the same time. Like fire departments, newsrooms rarely have all the people they want to cover the news and, being competitive, they don't rely on mutual aid. If you've scheduled your news conference for 10 a.m. Tuesday and the town's largest factory announces at 9:30 that it's closing forever, or if a double homicide takes place that same morning, few reporters may show up for your event.

One of the advantages of a news conference can also work to your disadvantage. By gathering all of the local media in one place at one time, you save yourself the trouble of having to do the same interviews over and over again throughout the day. The disadvantage is that there can be a certain synergism in a group of reporters, collectively asking follow-up questions and piggybacking off each others' inquiries in ways that can raise the emotional ante on a controversial story. If the issue at hand is controversial, and if you rarely hold news conferences, you may find yourself making the issue bigger just by calling in the press.

The occasions when a news conference works best are when your primary spokesperson is busy and doesn't have time to be available for interviews throughout the day; when your story has two talking heads, such as the chief and the mayor, and there's only one time all day when they can be scheduled together; or when you're demonstrating something, such as a new piece of fire apparatus, and can only do it once. Otherwise, you're often better off just sending out a news release and taking requests for interviews as they come.

If you've decided to do a news conference, you must address several planning considerations. The time of day is important, and you should take into account the deadlines, publication, and newscast times of the media with which you're working. For example, most television news operations try to get their crews back into the newsroom by late afternoon to assemble the evening's newscast. An evening paper needs to have the story put together by early afternoon to make the next edition. Part of the PIO's job is knowing the media's needs in his community and scheduling accordingly.

The day of the week can also be a factor. Monday mornings are usually slow news periods and, in most small and midsize cities, there is normally little major news on the weekends. If you're willing to work on a Saturday or Sunday, you may get more coverage than you would on a regular working day when your story is competing for space and time with more news items. Certain times of the year are notoriously slow—for example, the period between Christmas and New Year's Day, when most elected bodies are in recess and schools are on holiday.

You must also choose the setting for your news conference. If your purpose is to state the case for smoke detector maintenance, a conference room may be convenient and have plenty of electrical outlets for cameras and lighting equipment, but that conference room wouldn't pack the same punch as making your plea in front of the scene of a two-fatality fire. Since many communities have more television stations than newspapers, it's important to think visually, and don't forget that newspapers take pictures, too. If the story's about building or closing a fire station, why not do it on-site?

If, because of weather or subject matter, you're going to hold a news conference in a city hall meeting room, you can still think visually. Maps, blown-up photos, the charred remains of a child's doll—these and any other appropriate visual aids will help tell your story.

In addition to visual aids, plan on having printed handouts available to reinforce your key points. If, for example, you're talking about fire fatalities, have copies of statistical reports available to hand out. If the story is about a proposed hazardous materials response team, it may be useful to have copies of equipment inventories or training curricula on hand. Reporters may occasionally take a remark out of context, but documents go a long way toward ensuring accuracy.

At the news conference, the speakers should sit at a

table or podium, with chairs set out in front for reporters. A podium or lectern is ideal, since it allows speakers to make their statements and field questions without having to pass microphones back and forth. Work with television camera crews and newspaper photographers to help them get the pictures they need without blocking the view of other conference participants. This will usually entail vantage points from the sides of the room or over the heads of seated reporters.

A news conference should have no more than two or three speakers, each of whom should be prepared to make a brief statement and not all say the same thing. Speakers need to know that, no matter what they say or how long they speak, as little as fifteen seconds' worth of their remarks will make it on the air. This is the much-maligned "sound bite" that so dominates political discourse these days. It isn't necessarily right, but it's reality. The more concisely and colorfully a speaker can make his point, the better the chance it has of surviving the editing process.

The agenda is simple: The speakers should be introduced or introduce themselves, make their statements in prearranged order, and open up the conference for questions. The length of the question-and-answer period is determined by how complex the story is and how much time the speakers have available. This can be a judgment call, but don't let things drag. One way to end the period gracefully, if it's truly the case, is to announce that "the chief has a meeting with the mayor in ten minutes." Bear in mind that reporters often want to talk to the speakers one on one after the formal conference is over. A reporter with a specific angle on a story may not want to share it with a roomful of competitors but instead prefer to ask a couple of questions privately in a hallway outside the room. It's up to the speaker's time and comfort level if he wants to accommodate such enterprising reporters. Be sure at the conclusion of the news conference to thank everyone for coming.

Keep track of which news organizations show up for the

event. If, for example, a local radio station can't arrange to send someone, they may still appreciate a call and an over-the-phone interview when the news conference is over. Documents can be faxed or hand-delivered.

There are other arenas in which to tell your story. Part of your media list should include notations on what talk show opportunities are provided by the local television and radio stations. The PIO should make it a point to view and listen to these periodically so as to determine whether they might be interested in doing segments on issues relevant to the department. Talk radio has seen rapid growth in recent times, both locally and nationally, and it has become an important factor in shaping public opinion.

Live call-in radio is perhaps the most challenging type of media interview you can ever do. It provides direct access between citizens and public officials, and the questions can be about anything. You may think you're there to talk about summer fire safety, but when Bob on Line Two calls up to complain about how paramedics handled a call to his house, you'd better be ready to respond. In some cases, you may be able to deal with these war stories. In others, you're better off asking the caller to contact you later so you can look into his complaint or concern.

Finally, don't overlook newspaper editorial pages, which offer several opportunities for communication. Get to know whoever writes the editorials for the newspaper. In smaller communities, it may be a single person. Larger papers often have an editorial board. The people who write editorials spend much of their time in meetings with the people they write editorials about. If you have a major issue, a project on which your department is proposing to spend a lot of public money, or an issue on the ballot, you should arrange a meeting with whoever writes the newspaper's editorials. These meetings are usually held in an office or conference room at the paper and are less formal than a straight news interview, but be prepared. Frequently the editorial writers will invite the reporters who cover your agency to sit in on the discussion. Do your homework and anticipate hard questions. Editorial writers make their living questioning politicians and other advocates on often-controversial issues before deciding what positions their papers will take. A good confidence builder is to think of the worst, most difficult questions you might be asked and determine in advance how you will handle them. This is where a PIO with a good working knowledge of journalism and public affairs can be invaluable to a chief by giving the boss a sneak preview of what questions to expect in the actual meeting.

Another opportunity, also available through the editorial page editor, is that of writing an op-ed piece, so called because it runs on the page opposite the editorial page. Such pieces normally run as guest columns and can be a great way to get your point across if the talent exists within the department to write such a piece and the editorial page editor grants you the space.

Don't overlook letters to the editor. They're read by many people and are often used by different groups as a venue to sway public opinion. Your department may be on the receiving end of negative letters to the editor from time to time, and you'll need to decide whether or not you want to respond to them. If the letters express emotion or opinion, you may not want to write responses. On the other hand, if the letters show a misunderstanding of a situation or contain information that simply isn't true, you should consider a response to set the record straight. No matter how angry a letter makes you, your response should always be calm in tone and a straightforward recitation of the facts that support your position.

Finally, consider what proactive steps you can take to improve your relationship with the news media and their understanding of your operation. Fire departments around the country have undertaken different initiatives to educate reporters about their work.

A number of departments have instituted media days—

Communicating With the Communicators 51

miniature fire academies in which reporters are provided training in basic fireground operations and given the opportunity to suit up and fight controlled fires in burn buildings or vacant structures. The Phoenix Fire Department puts on a detailed, multiday academy for news people, at the conclusion of which they are certified as "fire journalists" and given passes authorizing liberal access to emergency scenes.

On a lesser scale, media ride-alongs are another way of building relationships and understanding. If you opt to allow ride-alongs, remember that it's not a perfect world. Consider carefully which of your crews you want the media to have extended contact with, and agree on guidelines for media access to private areas into which calls may take them. (See Chapter Five.)

Several departments, including my own, have created media pager programs in which newsrooms are given fire department pagers that can be activated to alert reporters and photographers to incidents in progress. Although there are some costs involved, such programs are a quick and efficient way to notify everyone equitably and fairly when incidents occur.

In smaller communities, newsrooms may be too short-staffed to allow them to attend events such as media academies, and a more passive approach may be in order. In the early 1980s, while working for the Eugene (OR) Fire Department, I wrote a guidebook for local reporters called *Working Fire,* which included explanations of basic firefighting procedure, descriptions of the different types of apparatus, translations of our radio codes, suggestions for fireground etiquette (don't drive over hose, don't park behind ladder trucks), safety information, descriptions of routine activities, key department contacts and phone numbers, and a glossary of common fire terminology. This guidebook proved popular enough with reporters that many took copies of it to jobs in other communities, and a number of departments modified the text for their own use.

There are, no doubt, other approaches that departments have taken to improve media relations. The key point is to be proactive and to talk to your local reporters about what works for them, not just wait until they call you.

Chapter Four

Who Can We Get to Do This Stuff? Choosing a PIO

The phone at headquarters rings, and an aide picks it up.

"Good morning, this is Sally at City Fire, how may I help you?"

"This is Dan Smith at Channel 5. We've just been told that the Metropolitan Women's Commission has called a news conference for three o'clock this afternoon at which they're going to charge that your department discriminates against women in its hiring practices. Obviously we want to give someone from the department a chance to respond. Can you help me?"

"Gee, I'm afraid you'll have to talk to Chief Kelly, and he's out of the office until tomorrow."

"Is there anyone else we can talk to? We're going to air the story on the Six O'Clock News this evening."

"I'm sorry, but Chief Kelly's orders are that all media calls go to him."

"Sally, can you tell me how many women firefighters you have in the department?"

"I'm sorry, I don't know if I can release that information."

News is just like fire—you never know when it's going to break out, and it always requires a quick response. Just as fire departments staff and prepare themselves to handle emergencies, so should they plan for public and media relations.

In this day and age, every department should have someone designated to fulfill this important function. How it's done depends on the department and the community. It can be the chief, if the chief is comfortable with the role and has the time to be available whenever the media comes knocking. It can be a volunteer who has good communication and interpersonal skills. If the need for public relations work isn't full time, the work can be done by a career firefighter or officer who serves as PIO in addition to his regular duties.

If the volume of work is such that PIO is a full-time position, there are several options to consider. Often a department will select a firefighter or officer who has the basic skills needed and appoint or assign him to the job. Other departments hire former reporters or public relations people and then provide them with training in the department and its work.

Many departments have public education officers (PEOs), who put on programs for school and community groups on fire prevention and life safety. A number of these departments use their education officers as information officers. This approach can work well, provided the department recognizes the difference between public education and information. The two are closely aligned and together make up a comprehensive public affairs program, but there are differences between the two functions. Public education is about prevention, giving community members the skills to prevent fires and other emergencies, as well as how to react properly when such incidents occur. Public information deals more with promoting the department's activities, providing information about incidents as they occur, and handling frankly political situations as they come up. If the workload allows, both tasks can be performed by the same person, as long as that person is skilled in both areas. In any event, it's important for PEOs to have good media skills, since the media provide an avenue for public education en masse. It's also important to note that assistance given to reporters covering incidents builds relationships and increases media willingness to do stories on prevention and life-safety topics.

It's also important to remember that everyone in the department has responsibility for relations with the media and the community. The main reason to appoint someone as PIO is to ensure that the day-to-day work of putting out information about incidents and other department activities gets handled, as well as to provide a friendly contact point for reporters seeking information about the organization. Like the rest of us, reporters prefer dealing with people rather than institutions. It's easier for both PIOs and reporters to work with people they know on a first-name basis. A lot of media relations is about human relations.

The chief plays a critical role no matter how the function is staffed. For the PIO to be effective, he must know what's going on in the department. It's best if the PIO reports directly to the chief and has regular access to the boss, as well as the ability to sit in on staff meetings. If PIOs aren't in the loop, they may release inaccurate information. Once this happens, reporters will lose respect and consider the PIO to be a barricade between themselves and the truth.

It also must be recognized that the chief must regularly talk to reporters, no matter how skilled the PIO. This is especially true in the case of major news, whether good or bad. It isn't necessary for the chief to release the basic information about a house fire, provide copies of fire statistics, or answer a reporter's questions about smoke detector maintenance, but when a multiple-fatality fire takes place or the department finds itself in a major controversy, reporters and the community expect the department's leadership to be visible.

Recognize also that PIOs shouldn't do all the talking for their departments, either. The best PIOs are like traffic cops, directing reporters to the best sources of information and prepping those sources to make the key points as clearly and concisely as possible.

PIOs should act as doors into their organizations, not walls. For example, if a firefighter effects the rescue of a child from a burning home, the PIO should instinctively know that the media will want to interview that firefighter and therefore should facilitate those interviews. The PIO may be the only one who has the full picture of the incident (i.e., times, number of companies and firefighters responding, damage extent and estimate, status of investigation, etc.), but it's important to get the people who actually fight the fires on camera as well.

With all this in mind, what skills and attributes should a department look for when appointing a PIO?

First and foremost, a PIO is a communicator. To this end, he needs good written and oral communication skills. Much of the work involves writing. All PIOs write news releases; many of them also write newsletters, articles, position papers, op-ed columns, and speeches. Oral communication skills are critical, especially given the

demands of emergency services, where PIOs must be able
to think on their feet and respond quickly.

Second, PIOs must be bright and able to be quick studies. The fire marshal may know more about the codes, the
paramedic more about heart dysrhythmias, the battalion
chief more about suppression tactics, and the chief more
about budgeting and strategic planning, but the PIO has
to develop a working knowledge about everything the
department does and, when the news focuses on a particular aspect of the department's work, the PIO must be able
to master the essentials of it quickly.

PIOs must have credibility. They carry the reputation of
their departments on their backs. In their contacts with
journalists and community members, they are the public
face of their organizations. They must tell the truth at all
times. There will be occasions when, for perfectly valid
reasons, the whole story can't be told about, say, an investigation or a personnel matter. The PIO must maintain
credibility by releasing whatever information he can and
articulately explaining why he can't disclose further information. Inside the department, PIOs must have credibility as well. They must be trusted by the people they work
with not to leak information—unless, of course, that's
what the department wants in a particular case.

Good PIOs have the ability to stay composed in the
worst of situations. In major emergencies, the PIO must
project a calm, controlled demeanor. He must also be a
skilled translator, able to take technical information and
explain it in terms the community can understand. The
ability to find the main points in nine pages of fire code
amendments and boil them down into a fifteen-second
sound bite that makes sense to the folks at home is a critical skill.

A good PIO must be something of a mediator—an
advocate for the department to the media and an advocate for the media within the department. One of the key
roles a PIO plays for the chief and management team is
the ability to serve as a sort of organizational conscience,

advising on how the media and the community are going to react to changes in policy and other issues that arise. Let's say, for example, that a town experiences a series of arson fires, causing widespread fear in the community. The PIO may well be caught between investigators who don't want to release any information and a community that is demanding to know what's being done about the situation. The PIO will typically be the middleman who convinces staff of the need to keep in touch with the community and who works with investigators to determine what he can release to the public and what he needs to withhold.

Finally, PIOs must be accessible twenty-four hours a day, seven days a week. News doesn't happen during regular office hours, especially in the fire service. The best PIO in the world is worthless to a reporter on deadline who can't get ahold of him. Consider the hypothetical example that begins this chapter—allegations of unfair hiring practices within the department. What kind of a story do you think Channel 5 is going to run on the news tonight? When can a PIO afford not to be on call?

A point to consider is whether a PIO wants the media to have home phone, pager, and cell phone numbers. I worked in a medium-size city that had a daily paper, three TV stations, and about twenty radio stations. I now work in the suburbs of a large city with even more newsrooms. I've always given the media my home and cell phone numbers, and the privilege has rarely been abused. In fact, there have been a number of times that it's enabled us to get our side across when controversial stories have broken on nights and weekends.

What does the rest of the department need to know to make the PIO successful? First, both management and the membership must understand that the PIO doesn't have sole responsibility for public relations. It can't be stated enough that everyone in the department has at least some responsibility for how the organization is regarded by the community.

Second, the members must understand the PIO's role and make sure that he is kept informed. Let's say that a fire company encounters a hostile property owner in the course of doing a fire inspection and the citizen threatens to go to the media. Someone from the company needs to contact the PIO to brief him on the situation. Similarly, if a crew is involved in a human interest story, such as delivering a baby under adverse circumstances, the PIO may be unaware of the event unless someone tells him about it.

Finally, members of the department should have a basic familiarity with information policy and be willing to talk to reporters themselves. Many departments have an internal ethic that it's somehow inappropriate to talk to journalists. Although certain topics, such as investigations and personnel matters, may be off limits, the average firefighter should feel free and able to describe what he saw on arrival at a fire building and what it felt like to go inside.

There's a certain irony in firefighters who mutter, "Damn, look who showed up," when the media arrive at a scene but who enjoy watching *Rescue 9-1-1* and *Emergency: On Scene* in their free time.

Reporters who are told that only the chief or the PIO can talk to them will come away with the sense that information is being overmanaged and so will be suspicious about the information they get. The best policy is one in which all department members are free to speak to the media but are responsible for what they say.

Despite their professional obligations, PIOs also take vacations, get sick, participate in conferences, and attend family events. A department needs to build some redundancy into its system to ensure that the basic work gets done when the PIO isn't available. It's like a relay race—someone should always be carrying the baton. The most critical work is incident-related, and additional members of the department should be trained to manage the on-scene presence of news crews and make sure information gets released within the parameters of department policy. Such redundancy can also prove useful when a major disaster occurs and a single PIO can't handle all the demands.

Finally, there are several pieces of equipment the department should invest in to support its public information program. The PIO needs clothing that is clearly marked and recognizable. This can mean either turnouts or some kind of jacket and cap. Many departments that have adopted the incident command system use vests to identify the key players. Any of these will work, as long as reporters and staff can quickly identify the PIO in the chaos of an incident.

Due to the nature of the job, a pager is critical. If the department has several people who are on call as a team, it's best to get multiple pagers, all with the same number. That way, the fire alarm office or communications center only needs to have one number on file for public information, and the on-call staff can spell each other without having to let the dispatcher know who's on call.

The PIO needs a scanner to monitor activity and a portable radio for responding to emergencies. While PIOs don't normally make a lot of on-air transmissions, it's critical for them to monitor the radio to know what's going on during incidents. It's especially useful when responding to a call to have some idea of what's going on while still en route.

A cellular telephone is an important tool. Suppose you're at the scene of a fire in a commercial building and, as the situation escalates, additional alarms are called and the streets around the fireground are closed to traffic. With a cell phone and a laminated wallet card listing media phone numbers, you can quickly contact local radio stations to alert motorists to the restrictions. Radio stations appreciate such live reporting, and the department thereby demonstrates its concern for the public.

There are number of other items PIOs should have available at incident scenes. These include lists of key phone numbers for the media, department staff, and local hospitals; reference materials on hazardous materials, such as the DOT guidebook; maps; and blank news release forms.

Chapter Five

Where Does It Say That? A Primer on Legal Considerations

The fire has been controlled, and the companies are in the process of picking up their lines. A photographer from Channel 12 stands in the front yard, shooting a firefighter rolling up a section of hose. Suddenly a large man steps out onto the front porch and bellows, "This is MY house, you son of a bitch! Get off my lawn!"

The PIO is in the office, and the scanner is carrying radio traffic on a routine medical call.

"General Hospital, this is Medic 5, we're en route to your facility with a nonemergency transport of a 29-year-old male patient. The patient's chief complaint is general weakness and respiratory distress. Our patient is confirmed HIV-positive. The vital signs are all within normal limits, we're administering O_2 by nasal cannula, and we anticipate arrival at your facility in ten minutes. Do you have any questions?"

"Medic 5, General copies your transmission. Can you provide the patient's name and his physician?"

Medic 5 transmits the name of a man known generally in the community as the mayor's son and the name of a prominent family practitioner. The PIO cringes as, within thirty seconds, his phone rings.

The fire at ACME Chemical has been controlled, but Building B has been sealed off as a hazardous materials hot zone, and members of a regional haz-mat response team are suiting up to enter the structure. The PIO is consulting with the incident commander when a reporter from a local TV station walks up.

"We want to know when this plant was last inspected, what violations were found, and exactly what chemicals are in that building," the reporter declares.

The environment in which PIOs operate isn't without rules. Public record and freedom of information laws, First Amendment rights, confidentiality of medical records, and bar/press guidelines are just a few of the legal areas of which skilled information officers should have a working knowledge.

What makes matters even more complicated, many of these laws vary from state to state and, given that they're periodically reinterpreted by courts and modified by legislative action, they're subject to change. Bearing all of this in mind, I offer a warning. I am not an attorney and, while this chapter will provide a basic overview of certain legal considerations, PIOs are well advised to research the laws in their own states, especially those pertaining to public records law. One of the best protections a department can have is a policy that spells out what information is releasable and what isn't and to have that policy reviewed by an attorney who is familiar with public records law. As the examples that begin this chapter show, it's important to know what the law is before trouble comes knocking.

A good PIO will make the effort to keep up with court

decisions that may affect his work. For extra credit, PIOs who have access to a law library are well ahead if they learn how to look up and read cases relevant to their craft. Many of the constitutional issues that govern media behavior aren't black and white. Those who look to the law will often be frustrated by the lack of absolute answers, but court decisions offer interpretations and can make interesting reading. A number of such stories relevant to the fire service will be told later in this chapter.

Most states now have public record (freedom of information) laws that spell out what kinds of documents and records are accessible to citizens and, by extension, journalists. The typical public records law divides information into three categories: information that must be released; information that may be released, depending on various factors; and information that is considered confidential and nonreleasable. It is vital to have an understanding of how these laws apply to the types of reports and information that fire departments handle on a daily basis.

In many parts of the country, fire incident reports are considered public information, and the information contained in them must be released to anyone who requests them, whether insurance adjuster, journalist, or any other interested party. (Note that most public records laws don't consider the motivations of people requesting information. If the record is public information, it must be released to anyone who requests it.) If, however, a fire is suspicious in origin, then the fire report becomes part of a criminal investigation and different rules apply. Although public record laws often require the release of basic information about crimes, they usually allow for withholding details that might compromise the integrity of a case. In the case of arson, for example, it can be useful for investigators to be the only ones who know exactly how the fire was started. Such information can be useful in evaluating tips, leads, and the occasional confession.

Another sensitive area that fire departments increasingly deal with is the medical information on patients they treat and transport. Unlike fire information, state and federal laws place restrictions on what can be released in medical situations, the theory being that prehospital care reports are every bit as confidential as the records in a doctor's office.

Other areas of public records law that PIOs need to be familiar with include the rules that govern personnel and disciplinary proceedings, inspection records, hazardous materials, and the conduct of public meetings, such as those of a fire district's board of directors.

It cannot be stressed enough how important it is for a fire department's release-of-information policy to be legally valid. Since the laws vary from state to state, it's important for readers to consult their own state's public records law or to seek legal counsel when developing policy.

Let's take a look at some of the major categories of information handled by fire departments and how laws generally apply:

Fire information is usually a matter of public record. In Oregon, for example, a fire report is considered public record unless the fire is arsonous. Even in those cases, the basic information (location, response and control times, and the number of personnel and apparatus) can usually be released without compromising the integrity of an investigation.

Fire departments must also deal with criminal investigatory information. Public records laws generally allow agencies to withhold information on criminal investigations but may not allow everything to be withheld. For example, the point of origin and mechanism of cause may be evidentiary, but that doesn't mean that other basic information about the fire isn't releasable. While some fire departments handle arson investigations alone, the more common model is for fire agencies to work in conjunction with law enforcement. It's important for fire officials working in such cooperative arrangements to coordinate the release of information closely with their law enforcement counterparts.

Another law enforcement situation that fire departments occasionally run into is the potential conflict between the First Amendment of the U.S. Constitution, which guarantees freedom of the press, and the Sixth Amendment, which guarantees the right to a fair trial. If a large amount of evidence about a crime becomes public record between an arrest and the subsequent trial, it can be reasonably argued that the defendant's Sixth Amendment rights have been violated. One of the more interesting findings of the Warren Commission report on the death of President John F. Kennedy was that news coverage of the assassination was so detailed and widespread that suspect Lee Harvey Oswald couldn't have had a fair trial anywhere in the United States. More recently, we have seen free press, fair trial questions raised in the O.J. Simpson case and in the media circus that surrounded a person of interest in the 1996 Atlanta Olympic bombing incident.

The main concern about pretrial publicity is that it can taint the pool from which jurors are selected. At the same time, journalists and some legal experts point out that there have been any number of high-profile cases in which it has still been possible to find groups of people so out of touch with current affairs that they are qualified to sit on a jury.

If a defense attorney can prove prejudicial pretrial publicity, there are several remedies the court can offer. One is continuance, in which a trial is delayed to allow publicity about the case to die down. The most expensive is change of venue, in which the trial is held in a different community from where the crime took place. This requires the prosecution and defense to travel, transport witnesses, and operate away from the resources that their own offices provide.

The American Bar Association has developed standards relating to pretrial publicity. These standards apply to law enforcement officers and attorneys, offering guidance on what should and shouldn't be released before a criminal trial. A number of states have adopted similar guidelines, often developed by committees made up of judges, prosecutors, defense attorneys, and representatives of the media. None of these bar/press guidelines, as they are often called, are binding on the media, but most law enforcement agencies have adopted them as policies on the release of information.

Medical information is often considered confidential due to privacy concerns. While there's probably no risk in reporting that a bike-riding child hit by a car suffered serious leg injuries, it's a whole different matter when a call involves AIDS, attempted suicide, alcohol, drugs, or mental problems. It is critical to establish policies for release of information that reflect the laws in your community. Such policies can serve as written explanation when questions arise about why certain information was or wasn't released.

For example, the attorney advising my former department interpreted Oregon law as saying that we had to release the identities of persons transported in our ambulances but that we could release no information as to their chief complaint or what treatment they received. Ironically, the local hospitals operated under a code of agreement between the medical community and the media that allowed them to release general descriptions of patient injury and condition but only to requesters who had the patient's name. A reporter seeking information on an accident victim would have to contact the police or fire department to get the patient's name and then contact the hospital to get the patient's condition.

Police cases can present unique challenges. There may be perfectly valid reasons not to release the identity of an assault victim, and investigators may want to withhold details of how a victim was assaulted and what injuries he sustained. I was involved in a case in which a woman was nearly killed by an assailant. The assault took place at a service station and was witnessed by several onlookers, clearly making it public knowledge and a news story. The man who assaulted the woman fled the scene and remained at large while she was being treated in a hospital. Investigators had reason to believe that the attacker might try to kill the victim while she was in the hospital. Accordingly, they didn't want to release her identity or even which hospital had admitted her. The media, while respecting the need of the police to withhold information, still had a legitimate interest in the woman's condition. By working with the public affairs staff at the hospital, we were able to issue periodic updates on the victim's condition without identifying her or what hospital she was in.

This example underscores another important point. It is very useful to have strong working relationships with information officers from other local institutions, such as law enforcement, utilities, hospitals, regulatory agencies, and schools. Situations invariably arise where the fire department is involved with these institutions, and it is easier to coordinate the release of information if you already know your counterpart in the other organization. This is especially critical in disasters and major incidents where multiple agencies are involved.

The scanner can be a source of concern when it comes to medical information. Most newsrooms monitor police and fire radio traffic and are therefore able to hear conversations between paramedics and receiving hospitals. This information can be useful when newsrooms make decisions such as whether to send a crew to cover a traffic accident. Severity of injury is one of the considerations an assignment editor might take into account in deciding whether to commit resources to a story. At the same time, sensitive information is occasionally transmitted over the radio. Many departments use codes to transmit such information, but it doesn't take long for scanner moles to figure out the codes. PIOs should have some knowledge of the Communications Act of 1934, a piece of federal legislation that provides that no unauthorized person shall receive, intercept, divulge, or publish the contents of any radio transmission. A number of states have similar laws on the books. It is generally agreed that the monitoring of radio traffic is not at issue but, rather, the interception and divulging that would occur if such information were published or broadcast. Enforcement of the federal law is left up to the Federal Communications Commission and the Justice Department and, given the workload and staffing of those agencies, probably isn't the highest priority. At the same time, would-be users of intercepted communications should be aware that they risk potential civil action for violating the privacy of patients in situations such as the HIV scenario given at the beginning of this chapter. In any event, information officers have no obligation to officially confirm information that third parties obtain by such means.

One of the trickiest issues is that of media access to emergency scenes. How close should photographers be allowed to get? On what grounds can reporters be kept out of an area? Can news people go onto private property? Can news photographers on ride-alongs go into homes?

The issue of media access is one of ongoing debate and evolving law. From the standpoint of the media, it's critical for them to get as close as possible to news events, and the growing show-and-tell nature of television makes getting closeups all the more important. From the standpoint of police and fire officials, there's no denying that the presence of the media complicates a scene, creating logistical demands ranging from the parking of satellite trucks to requests for interviews. The matter is further complicated by how the two parties view themselves. Public safety officials often see themselves as having ultimate authority over emergency scenes, while news organizations cloak themselves in the First Amendment and remind us that they rode helicopters with the troops in Vietnam.

The best guidance for public safety officials working with the media at incident scenes is to recognize the legitimate role of the media in covering such events and to provide reasonable accommodation. Ideally, reporters shouldn't be considered the same as civilian bystanders but, rather, given status similar to the crews that come to shut off the utilities to a burning building. They shouldn't get as close as emergency responders but shouldn't be kept back across the street, either.

To begin with the most basic concept, the media has the right to be present and take pictures in public areas. If an accident occurs in a busy intersection, the media has the right to be at least as close as the general public and to take pictures. Fire departments that want to be responsive to the media's needs often establish two perimeters at incidents: one for the general public and a second, closer one for the media.

As this is being written, only two states have laws on the books that specifically address the issue of media access to emergency scenes: California and Ohio. California Penal Code Section 409.5 provides in part: "(a) Whenever a menace to the public health or public safety is created by a calamity such as a flood, storm, fire, earthquake, explosion, accident, or other disaster, officers of the California Highway Patrol, police departments, marshal's office or sheriff's office ... or designated peace officer may close the area where the menace exists for the duration thereof by means of ropes, markers, barricades, or guards ... (b) ... officials may close the immediate area surrounding any emergency field command post or any other command post activity for the purpose of abating any calamity ... (c) any unauthorized person who willfully and knowingly enters an area closed pursuant to this section and who willfully remains within the area after receiving notice to leave shall be guilty of a misdemeanor ... (d) Nothing in this section shall prevent a 'duly authorized' representative of any news service, newspaper, or radio or television station or network from entering the areas closed pursuant to this section."

There has been only one court case interpreting this legislation. In that case, a California appeals court opined that "Press representatives must be given unrestricted access to disaster sites unless police personnel at the scene reasonably determine that such unrestricted access will interfere with emergency operations. If such a determination is made, the restrictions on media access may be imposed for only so long, and only to such extent, as is necessary to prevent actual interference."

The court went on the say that "Safety is not a reason to exclude press members from a disaster site because the statute 409.5(d) provides a specific exception for members of the media, in situations already determined to be unsafe. This means that members of the press must be accommodated with whatever limited access to the site may be afforded without interference." (Leiserson v. City of San Diego (1986) 229 CR 22.)

Ohio Penal Code Section 2917.13, Misconduct at an Emergency, states: (A) "No person shall knowingly: (1) Hamper the lawful operations of any law enforcement officer, fireman, rescuer, medical person, or other authorized person, engaged in his duties at the scene of a fire, accident, disaster, riot, or emergency of any kind; (2) Fail to obey the lawful order of any law enforcement officer engaged in his/her duties at the scene of or in connection with a fire, accident, disaster, riot, or emergency of any kind. (B) *Nothing in this section shall be construed to limit access or deny information to any news media representative in the lawful exercise of his duties.* (C) Whoever violates this section is guilty of misconduct at an emergency, a minor misdemeanor. If violation of this section creates a risk of physical harm to persons or property, misconduct at an emergency is a misdemeanor of the fourth degree."

The Ohio code, while providing a legal means to control bystanders at emergency scenes, clearly recognizes the role of the media in covering such events.

Outside of the two statutes referenced above, reporters and photographers generally have no more right of access to emergency scenes than the general public, but fire departments wishing to maintain positive relations with the media will accommodate the media as much as possible.

What about other restrictions on media access? While the California law says that safety is not a consideration, most other public safety organizations do take media safety into account when drawing lines for access. An argument could be made that a journalist allowed into an unsafe area would, if injured, require resources for his care that could otherwise be deployed in mitigating the original incident. At the same time, reporters and photographers can get pretty worked up when told they can't enter an area because they lack the proper safety clothing, only to see shirt-sleeved, unhelmeted firefighters and officers milling around inside the line. If you establish those kinds of restrictions at a scene, be vigilant about what your own people are wearing inside the line.

Another general restriction, also recognized in California, is the preservation of a crime scene. PIOs need to remember that any accident or fire scene is a potential crime scene until proved otherwise. Investigators have a legitimate interest in limiting the number of people who walk over evidence before it can be documented and collected. In addition, a case can be compromised by the publication of photos or descriptions of evidence found at the scene. For example, it's already been pointed out that it's useful to have some fire cause and origin information known only to investigators, enabling them to evaluate tips and leads. If cause and origin information is disseminated by the media, it makes it more difficult for investigators to evaluate the information they receive from other sources.

The same California appeals court that interpreted that state's law on media access noted that "Media members do not have any access rights to any immediate crime investigation, or police SWAT team locations, when public safety officers or investigators believe a crime has been committed."

What constitutes actual interference at an emergency scene? Courts have made it clear that interference must be more than a matter of responders being annoyed by the presence of cameras.

On the afternoon of March 25, 1977, Harvey Lashinsky, a photographer for the *Newark Star-Ledger*, was driving south on the Garden State Parkway in New Jersey when he saw a broken section of guardrail and an overturned car down the embankment. Lashinsky parked his car about 150 feet from the scene and walked back to take pictures of the accident.

Some 15 to 20 minutes later, State Trooper Eric Herkloz arrived at the scene. By this time, a crowd of some forty to fifty onlookers had stopped to watch, and a volunteer first aid squad member who had stopped to assist told Herkloz that the woman driving the car was deceased. Pinned against her was a girl in serious condition and going into shock. Gas, oil, and transmission fluid were leaking from the vehicle, and personal property was littered about the scene. The car's battery, still attached by its cables, had fallen from the car and cracked open.

Herkloz, the sole responding officer, called for additional troopers and an ambulance. Concerned about the potential for the vehicle to catch fire, and wishing to preserve evidence, Herkloz ordered everyone back from the scene, with the exception of two persons who were rendering first aid to the injured girl. Lashinsky remained taking pictures and was asked, individually, to "please leave the scene." The photographer stepped back about five feet and moved no further, producing a press pass issued by the New Jersey State Police. The trooper told Lashinsky, "I don't care at this point" and again asked him to "please leave the scene."

At this point, accounts differ. At his trial, Lashinsky testified that Herkloz arrested him immediately. Herkloz and the two people rendering first aid, however, testified that the photographer engaged the trooper in a heated argument lasting three to four minutes, during the course of which Lashinsky swore at Herkloz and told the trooper to go do his job and to let Lashinsky do his. Lashinsky was arrested on a disorderly persons charge.

At trial, it was argued that the photographer had impeded the officer's ability to perform his duties. The size of the crowd made it difficult for a lone officer to exercise control. When the argument broke out between the two men, a number of the bystanders who had been moving away from the scene stopped to watch the dispute. Herkloz testified that he wanted to assist with first aid, and a woman who was providing first aid testified that she wanted help from the trooper but couldn't get his attention because his back was turned to her and he was arguing with the photographer.

Lashinsky's defense at trial was that he didn't directly, physically interfere with the trooper and that he had no intent to interfere with the officer. A lower court failed to see a compelling argument in Lashinsky's favor and he was convicted of disorderly conduct.

What should be of interest to PIOs is what happened when Lashinsky's case was appealed to the Supreme Court of New Jersey. The court upheld the arrest, but only by a four to three vote. The dissenting justices emphasized the importance they placed on the right of journalists to gather news, noting that the state police had recognized that right when they issued press cards identifying possessors as responsible individuals engaged in a task deeply affected with the public interest, and thus "as an individual who in the discretion of (a) police officer can proceed beyond that point where the public goes if it fits in with what is going on at the time."

Noting that journalists are still required to obey the reasonable commands of police officers, the dissenting judges questioned the reasonableness of Herkloz's command for Lashinsky to move back. Lashinsky was not a mere bystander—he was a news photographer engaged in the task of recording a newsworthy event. Standing fifteen to twenty feet away from the wreck, he was in no way destroying evidence or interfering with first aid efforts. (State v. Lashinsky, 404 A 2d 1121.)

Although Lashinsky's arrest was upheld, the narrow supreme court vote indicates the difficulty that even judges with ample time for consideration have in deciding access issues, underscoring the importance of reasonably accommodating the media at incident scenes.

A similar case took place in California. Shortly before midnight on July 6, 1984, a traffic accident occurred in which a Mazda RX-7 wound up wedged under the rear of a tractor-trailer rig. City of Whittier police officers and a private-sector ambulance responded. Noting the need for extrication of the female driver of the car, Stuart Donald Cumming, one of the two paramedics on the private ambulance, called for help from the Los Angeles County Fire Department. As firefighters began using a hydraulic device to free the woman from the wreckage, Cumming, his partner, and fire department paramedics began treating the woman, who appeared seriously injured.

As the rescue effort was taking place, two photographers, Patricia Ruth Olsen and her husband, showed up at the scene. Ms. Olsen would later testify that she was a professional photographer who often took photos of fires, accidents, and other incidents and would submit them to emergency services publications.

As the paramedics worked on the victim, Olsen began taking pictures using a flash or a strobe light. Although she was warned by a police officer to stay out of the way, when the injured woman was placed on a gurney, Olsen leaned over the gurney to take pictures and took a series of photos from as close as two to three feet.

Cumming later testified that he was concerned because the flash was leaving blind spots in his vision and that the patient was moving and jerking in apparent response to the light. Further, both Cumming and a police sergeant at the scene questioned the appropriateness of such close-up pictures, since the patient's blouse was open.

Cumming repeatedly asked Olsen to move back, at times directly and at other times through the police officers. Olsen was defiant, asserting that she had a right to be there to take pictures and, at one point, asking Cumming his name and company so she could file a complaint against him. On at least two occasions, Olsen extended her wrists to police officers and told them to arrest her, but they declined, saying they were busy and just wanted her to cooperate.

Finally, as the gurney was wheeled to the ambulance, Olsen moved to within a foot of it, leaned over a kneeling paramedic's shoulders, and began shooting photos of the patient using the flash. At this point, Olsen was arrested and charged with interfering with the lawful efforts of a fireman, firemen, or emergency rescue personnel in the discharge of official duty and disobeying the lawful order of a fireman or public officer.

In a jury trial in the Whittier Municipal Court, Olsen was convicted, but the case took a turn on appeal. Olsen's arrest was made in response to requests from Cumming, a private sector paramedic. An appeals court held that the language of the law under which Olsen was arrested made it clear that the statute applied only to interference with and failure to obey the orders of public sector emergency responders. The court reversed Olsen's conviction, noting that private sector paramedics were deserving of the same protection and suggesting "that the Legislature exercise its will to that end." (People v. Olsen, 230 Cal. Rpt. 598.)

The two cases described above are just examples of the kinds of conflicts that can arise between responders and the media. Many similar incidents have taken place but have been resolved out of court when public safety and news organizations have recognized the need to maintain cooperative, long-term relationships and have negotiated their own settlements. It's important to work with your local media to reach a mutual understanding of what access and reasonable accommodation mean.

While it's generally agreed that the media have the right to be present at emergency scenes in public areas, private property raises some complications. Consent to enter private property is between the media and whoever has control of the property—not public safety responders. Many fire departments, for example, protect large industrial plants and complexes. If an emergency occurs in one of these, the management of the plant has control over whether or not reporters are allowed onto the premises.

The situation gets even stickier when incidents occur

on private property but the person having control of the premises isn't on hand to grant or deny permission to enter. Fire departments cannot grant permission for reporters to enter, nor have they been held responsible for keeping the media out.

On the afternoon of September 15, 1972, a 17-year-old girl died in a house fire in Jacksonville, Florida. The girl's mother was out of town, visiting a friend. Fire and police investigators invited the news media to enter the house, as was standard practice at the time. The fire marshal wanted a photo showing the silhouette that had been left on the floor after the victim's body had been removed. Out of film and unsatisfied with the single picture he'd taken with an instant-picture camera, he asked a newspaper photographer to take the silhouette photo, which was subsequently made a part of the case file.

The news photographer also turned the pictures over to his paper, and the silhouette photo was published in the September 16 edition of the *Florida Times-Union*. Klenna Ann Fletcher first learned of the facts surrounding her daughter's death by reading the newspaper story and viewing the published photographs. She sued the newspaper, claiming trespass and invasion of privacy.

The case eventually found its way to the Florida Supreme Court, which found that, since it was customary for public safety officials to invite the media onto private property to cover newsworthy events, and since no one was at the home to object to the media's entry, there was no trespass or invasion of privacy. The case was appealed to the U.S. Supreme Court, which declined to hear it. [Florida Publishing v. Fletcher, 340 SO 2d 914 (cert. d. 431 US 930.)]

Although news organizations were happy with the outcome of the Florida case, a number of subsequent law enforcement cases changed the picture considerably. In several of them, the courts found that law enforcement investigators could not bring the media onto private property in the course of raids and investigations. It should be noted that most of these were civil actions brought against news organizations by aggrieved property owners. The public agencies involved were usually not named in the lawsuits.

On the night of October 30, 1979, Dave and Brownie Miller were at home in their Los Angeles apartment when Dave Miller suffered a heart attack while in his bedroom. Police and L.A. City Fire Department paramedics responded to the apartment, where it was determined that Mr. Miller was in cardiac arrest. Mrs. Miller was escorted from the bedroom as paramedics initiated CPR and other resuscitative measures.

But what Brownie Miller didn't know at the time, and would dramatically find out later, was that paramedics weren't the only people in the bedroom with her dying husband. Emergency medical services in Los Angeles, provided by a mix of public and private agencies, had become a source of community concern, and KNBC-TV had decided to produce a week-long series of news stories about the issue. As part of its coverage of the story, KNBC had approached the L.A. City Fire Department with a request to have a television crew ride along with fire department paramedics to get video of the responders in action. Fire officials, eager to cooperate with the media, agreed to the ride-alongs. There was no discussion by either television producers or fire officials about getting permission before entering homes. By the time Dave Miller had his heart attack, the KNBC crew had accompanied fire department paramedics into some ten to fifteen private residences. The producer of the series, who accompanied his crew on the ride-alongs, later testified that on about half of these occasions, someone in the home would ask what the reporters were doing but didn't object once they'd been told.

Dave Miller was pronounced dead after he arrived at a local hospital. Some three weeks later, his daughter was watching the news. One of the stories about EMS was airing, and she was shocked to recognize the inside of her parent's apartment. Although the victim's face wasn't shown, she recognized tattoos on his arms as being her father's. She called the TV station to complain and then called her mother. As a result of her daughter's warning, Brownie Miller didn't watch the news all week but, while looking for a soap opera one midmorning, came upon a promotional spot that showed video of her home and her dying husband. She, too, called KNBC to complain.

In May 1980, a lawsuit was filed by Brownie Miller and her daughter against the television station and the City of Los Angeles, claiming trespass, invasion of privacy, and emotional distress. The suit asked for a half-million dollars in damages.

The case moved slowly. In September 1984, the City of Los Angeles moved for immunity and was released from the lawsuit. But the legal wrangling between the Millers and KNBC continued for nearly five years. On April 24, 1985, a California trial court dismissed all of the plaintiff's claims.

The family appealed, and in December 1986, the Court of Appeals ruled that the television crew had clearly intruded into Brownie Miller's private space and that she could proceed to trial. It had taken six years just to get to the point where a trial could begin. The parties settled without going to trial, and the amount of compensation given to Mrs. Miller has been kept confidential, as stipulated by the parties in the settlement agreement. (Miller v. NBC, et al, 232 Cal. Rpt. 668.)

Media ride-alongs have become more common, especially in law enforcement vehicles, and a whole cottage industry of reality-based television shows has arisen around such activities. While many of these programs use reenactments not subject to invasion of privacy claims, a number do use video obtained while accompanying responders.

In March 1992, federal agents invited a network television crew to join them as they served a search warrant on a New York apartment in the course of a credit card fraud investigation. The suspect was not home, but his wife and young son were. The television crew videotaped agents as they searched the apartment and questioned the wife. The suspect was later indicted, but his wife sued on behalf of herself and her son, claiming their privacy had been invaded. A federal judge ruled that the case could proceed to trial, noting that the search warrant gave government agents permission to enter and search the apartment but did not give them the right to bring along the TV crew, noting that "CBS had no greater right than that of a thief to be in the home...." Significantly, the judge ruled that the plaintiff had cause for action not only against the media, but also against the public officials who facilitated the media's involvement in the first place. (Ayeni v. CBS, Inc. 848 F. Supp. 362.)

To sum up, the media access issue is a balancing act involving First Amendment protections of newsgathering activities, the legitimate needs of public safety agencies to do their work and to preserve evidence, and the privacy rights of those who may unwittingly find themselves in the news.

The safest course for any fire department is to develop policies that recognize the right of news organizations to respond to and cover incidents, to provide reasonable accommodation and access at emergency scenes, and to be wary of cooperating in media initiatives that could violate someone's privacy. As noted in the cases presented here, this is an evolving area of the law, and fire departments are well advised to keep up with court decisions that could modify or change the rules of engagement.

Chapter Six

The PIO at the Incident Scene

It is 5:30 a.m., and a northbound freight train is only partially onto the siding when a southbound freight slams into it. With a thunderclap of impact and a screech of metal on metal, the six locomotives of the southbound train tear into several boxcars, derailing themselves and twenty cars in the process, sending wheels, cargo, and a half-ton of earth and rock skyward. The sound of the impact echoes across the flat land, awakening farmers. Within seconds, the crash is over and silence returns to the surrounding fields as a plume of smoke, thick and black, spirals from liberated diesel fuel. Two area residents, awakened by the sound, look out their windows and call the regional 911 center. Within a minute, several volunteer fire companies are dispatched to a "train wreck, with fire, unknown if injury."

Some twenty miles away, finishing her first cup of coffee, a reporter for a local radio station hears the dispatch on a newsroom scanner. Adrenaline flowing, she calls the 911 center to confirm the information and is told that multiple callers have reported a collision between two freight trains in an isolated rural area. With no more information than this, she goes on the air with an unconfirmed report that a major railroad accident has occurred. After alerting her listeners, she calls the regional office of the Associated Press.

Before the first emergency crew has arrived on the site, the media has begun to cover the incident. As fire engines, medic units, and police vehicles converge on the scene, a news director in a television station 120 miles away makes the decision to send a crew with the station's satellite truck, and his counterpart across town calls up his station's helicopter.

The most dramatic and publicly visible part of a PIO's job is responding to emergency incidents, managing the on-scene presence of the news media, and serving as spokesperson for the department.

What is the work of a PIO at an incident? At any emergency scene, media relations work falls into two categories: access and information. Access is all about how close photographers, still and video, can get to the scene to get the pictures they need. Information, of course, deals with how much detail your agency will gather and disseminate to reporters covering the incident.

To begin with the most basic question, how do newsrooms know when an incident is occurring? Most newsrooms monitor scanners to follow police and fire activity but may not always know when an incident is taking place. In midsize and smaller communities, for example, newsrooms often aren't staffed during the late night and early morning hours. Reporters might not know there's a two-alarm fire at the high school unless someone calls and tells them about it. There's certainly no requirement to do so, but a fire department can certainly strengthen its relationship with the media by maintaining a list of the home phone numbers of key news people and calling them when major news occurs in the middle of the night. Some departments, including my own, have issued pagers to local news organizations so they can quickly be alerted when major incidents take place.

Another issue to consider is what the rules are for dispatchers in providing information to reporters. Some busy communications centers won't take media calls, period. Others allow dispatchers to confirm the basics, such as "we have a working house fire at 1122 North 30th Avenue." From a PIO's perspective—especially a PIO who doesn't want to field media calls at home every night that there's a false alarm at the hospital—the latter policy is preferable.

The department also needs to establish policy as to what level of incident the PIO will be alerted to. This will vary from community to community. In some cities, any working fire is considered newsworthy and a PIO will be dispatched. In larger cities, fires need to go to multiple alarms, involve fatalities, or result in multiple injuries to be newsworthy and get a PIO to the scene. Some departments have protocols that allow chief or company officers to serve as PIOs at smaller incidents. This system works well as long as these officers have a basic understanding of what information is and is not releasable.

Departments should recognize that reporters may show up before the PIO and should make provisions to release basic information. Too often, firefighters are overly paranoid about talking to reporters. Good departments ensure that all members are aware of the policies for releasing information, and incident commanders should have the discretion to talk to reporters, or assign someone to manage media presence, before the PIO arrives. It shouldn't be a state secret to release basic information—for example, that firefighters are searching a structure for victims or trying to prevent a fire from spreading to a neighboring building.

There's a classic anecdote in public relations circles

about a refinery, owned by a large corporation, with a policy that information could only be released to the media after being reviewed and approved by the plant manager. One night, the plant manager went to play poker at a friend's house and forgot his pager. An explosion and fire occurred at the refinery, and the plant's public relations man was seen on the Eleven O'Clock News pacing back and forth in front of the main gate, saying, "Fire? What fire?" He'd been unable to receive clearance to confirm what could have been deduced by anyone driving by the plant. To provide a fire service corollary, nothing annoys reporters more than being told that they'll have to wait for a statement from the PIO, only to see the PIO arrive and require a briefing from the same officials who wouldn't talk to them.

Let's assume that an incident has occurred requiring the response of a PIO. What are the steps the PIO should take in handling the call?

The first step is to get there, hopefully monitoring the radio while en route to learn as much as possible about what's going on. Depending on the time of day, the PIO may want to call newsrooms to make sure they're aware of the incident. If the incident has resulted in street closures, the PIO should notify local radio stations as soon as possible. If the PIO has a cellular telephone, he can accomplish these tasks while en route to the scene. It's useful to carry a laminated wallet card containing the day and night phone numbers of local news organizations and their key personnel.

On arrival, the PIO should don either marked turnouts and helmet or an ICS-style vest so that reporters and photographers can easily identify him. The PIO should then report to the command post. Under the incident command system, the PIO reports directly to the incident commander so that he can receive a quick briefing on the status of the incident. Part of this briefing should cover whether other agencies are involved, which could indicate the need to share information and coordinate work with other PIOs.

For the average house fire, this isn't normally an issue, but it becomes one when dealing with events such as major transportation accidents, hazardous materials incidents, rural/urban interface fires, and criminal activity.

Once briefed, the PIO should move away from the command post. The role of a PIO in an emergency is to handle on-scene media relations by "taking the heat" off the incident commander. There is enough stress, tension, and controlled chaos at the command post without having cameras and microphones looking over everyone's shoulders. The PIO should set up near enough to the command post to stay in touch but far enough away so as not to bring the media into the command area.

In addition, the PIO should quickly survey the incident scene to determine what safety hazards exist or what evidence must be preserved, as well as where media people can and cannot go. As noted in Chapter Five, the media generally do not have a greater right of access than the general public. At the same time, a department that wants to maintain a good relationship by recognizing the media's role will see that reporters have better access than the public at large.

All of this is easier if barrier tape has been placed to define the fire scene. Within this outer perimeter, it may be possible to establish a media area, allowing photographers to get unobstructed close-in shots while keeping them out of the way of department operations. If, for example, a fire is on private property, and the owners or managers of the property object to the presence of the media, it may be necessary to restrict the reporters to public areas. This is sometimes the case when emergencies occur in large industrial plants. If the incident is on private property, the owner has control over who is allowed on the premises.

Resolving access issues can require creativity on the PIO's part, but efforts to get the media as close as safely and legally possible will go a long way toward maintaining a positive relationship.

Rules and restrictions about access must be conveyed to news crews as they arrive and be strictly enforced by the PIO. Newsgathering is a highly competitive business, and if Channel 6 has been told they can't go into the backyard, they'll be very upset if Channel 4 gets video from there. (At the same time, cameras don't always present problems. The bright lights on television cameras often help to illuminate night scenes.)

Another issue that arises in some situations is media access to the airspace over an incident. The helicopter has become a major newsgathering tool in many cities. Although aerial images of incidents such as wildfires and major freeway accidents make for good television and can be of great value to incident commanders, airborne news coverage can also pose problems. Concerns have been raised when major incidents bring so many news and public safety aircraft into the area that serious potential exists for a midair collision. Helicopters provide stable platforms for gathering video, but they can make ground communications difficult and, in the worst cases, rotor downwash can push smoke or toxic clouds back onto responders and bystanders.

If a fire agency has its own air unit, the coordination of airspace will probably be done by a designated air operations coordinator, as is usually the case with major wildfires. Most media pilots are responsible and will contact public safety aircraft or the air coordinator if they know fire or medical units are flying on the incident. Still, any PIO who works with newsrooms that use aircraft needs to develop, in advance, an understanding of how to communicate with a news aircraft that shows up to cover an incident. Typically this will be through the newsroom itself, which stays in radio contact with the aircraft.

For worst-case scenarios, PIOs should be familiar with Federal Aviation Regulation 91.137, which allows for temporary flight restrictions over hazardous areas. A temporary flight restriction establishes a zone over an incident scene and limits air traffic within that zone. This is accomplished by the issuance of a Notice to Airmen (NOTAM) by the Federal Aviation Administration. The process is initiated by notifying the nearest FAA office, which must then share the information with regional and national centers before the NOTAM can actually be issued and transmitted—something that can take several hours to accomplish.

Another key point that must be recognized is that fire officials aren't responsible for the images that news organizations choose to print or broadcast. Periodically, you will read reports about how firefighters or other emergency responders have kept the media away from a scene or put their hands over lenses in the belief that the pictures they're taking aren't fit for public consumption. Again, media personnel have a legal right to be at emergency scenes and take pictures. It is ultimately an editorial decision as to what images are used to report a story.

In July 1975, Stanley Foreman, a photographer for the *Boston Herald-American*, went to a fire in a multistory brick apartment building. As Foreman watched, a dramatic scene developed. A woman, a child, and a firefighter stood on the top level of a fire escape as an aerial ladder was raised to their position. As the ladder approached the fire escape, the firefighter reached out to grab it just as the fire escape pulled away from the face of the building. Using a motor-driven camera, Foreman shot a dramatic series of pictures as the firefighter clung to the underside of the ladder and the woman and child fell to the sidewalk below. The pictures he shot weren't only published in the *Herald-American*, they were also put on the Associated Press wire and run in papers across the country. Predictably, as is often the case when newspapers run graphic photos, there were outraged calls to editors and a number of canceled subscriptions. What's equally important to note is that no one blamed the Boston police or fire department for allowing Foreman to take the pictures, which eventually won a Pulitzer Prize.

We are not the "taste police." The things we see and deal with are often grim, but it isn't our responsibility to shield the rest of the world from seeing them. Most news organizations have their own standards for what they will air or print, and they often shoot images that they will never use. It's not uncommon for a photographer in the middle of an emotionally charged incident to take pictures of everything, then later decide in conjunction with news directors or editors what will get published or broadcast.

The only conceivable situation in which we have a legitimate objection to what's photographed at a scene is if it is disturbing a patient or victim to the point that it compromises our ability to care for or interview him. Even in those situations, our objections are usually stronger on moral grounds than they are on legal ones.

While dealing with access issues, the PIO must also begin gathering information. Using the forms described in Chapter Three, the PIO should begin to assemble information about the incident and prepare to release it to the reporters at the scene. Most incident command protocols say that the PIO should clear any information with the incident commander before turning it over to the media. In reality, the need for clearance depends on the skill of the PIO and his working relationship with the incident commander. It is advisable to review the information with the IC for accuracy and to share any information that the PIO himself has learned of which the IC may not be aware.

If the PIO needs time to gather information and reporters are at the scene, an approach that works well is to tell them that you'll meet them at a specific location and time, taking into consideration the deadlines of the reporters with whom you're working. "Go ahead and talk to the witnesses. Let's meet by that hydrant on the corner in ten minutes and I'll tell you whatever I can find out." If the incident occurs on deadline and the reporters have to get back to their newsrooms, take their phone numbers and make the commitment to call in whatever information you can get before the deadline passes.

Another consideration is those newsrooms that haven't sent anyone to the scene. As discussed earlier, many radio stations operate with limited staffing and only gather news over the phone. If this is the case in your community, don't forget to give them a call. Before I was issued a cellular telephone, I used to carry a roll of quarters in my car to cover such situations.

It's also important to see that all reporters get equal treatment. The second most grievous sin a PIO can be accused of, after lying, is playing favorites with newspeople. At incidents, it's important to see that everyone gets the same basic information and access. The reporter from Channel 3 may be a former Miss Nebraska and the guy from the newspaper may look like your Uncle Ralph from Brooklyn, but they're both going to get the same level of assistance and courtesy from you at the scene. Besides, the newspaper's probably going to have more information about the job in tomorrow's edition than Channel 3 will at 11 tonight.

Once the situation has been controlled, the PIO has several options. If the reporters have left the scene, the PIO may head back to headquarters and tend to whatever requirements local procedure dictates. This may include putting a news release on an answering machine or faxing it as discussed in Chapter Three, then routing copies to other divisions of the department that need to know what's happened. At minimum, this should include the chief's office and the fire prevention bureau. A copy of the release needs to be filed in the public information office for future reference.

What if you don't have the whole story? It's not uncommon for a fire to be controlled in thirty minutes and for the investigation to take hours or days. Depending on the time of day, workload, and media interest, the PIO must decide whether to stay at the scene with investigators or leave. If the PIO leaves, he must make arrangements to stay in touch with investigators.

In any event, information relating to fire cause determination is often sensitive. If it's obvious that a fire resulted from an accidental cause, information about that cause can be released. If, in fact, the fire resulted from some unsafe behavior or careless act, the PIO may have an opportunity to get across, via the media, some key messages on fire safety.

In some cases, investigators may be unable to get any further than a possible or probable cause. If such is the case, the information should be clearly released as possible or probable. It's not uncommon for fires, especially those with major losses, to be subsequently investigated by insurance or other investigators working on civil litigation. A PIO who has released a tentative cause as a confirmed fact may be embarrassed later.

In any event, a PIO must work closely with investigators, especially in the case of suspicious fires. Investigators must ultimately have control over what information gets released about any fire involving possible arson. Depending on how an agency handles fire investigations, the PIO may be working with fire marshals, police detectives, or a joint team.

A general policy followed by many departments is to acknowledge that a fire is suspected arson but to withhold specific details on point of origin and mechanism of cause. Saying that a fire is arson may produce helpful information from witnesses or informants, especially if there is a local or state reward program as exists in many areas. At the same time, it is useful in any investigation to have a body of information known only to investigators and whoever committed the crime. This information can help determine the validity of any anonymous tips, leads, or confessions. Withholding specifics can also serve to prevent copycat firesetting. Finally, many departments have a philosophical objection to telling people how to set fires. The bottom line is that decisions about the release of information require careful thought and ultimately rest with those investigating the fire.

In the early morning hours of June 28, 1996, a fire in an Aloha, Oregon, apartment building took the lives of eight people. A team of local, state, and federal investigators determined that the fire had been deliberately set in a stairwell that was the sole exit for twelve apartments. Reporters were told that the fire was suspected arson and that it had begun in the stairway, but no details were released regarding the number and points of origin or mechanism of cause.

Several nights after the fire, an engine company responded to a pull-station alarm in a hotel near the fire scene. The company experienced some difficulty in getting the alarm reset and radioed a call for assistance. An on-call deputy fire marshal, one of the principal investigators on the apartment fire, responded and assisted the company in resetting the alarm. With the alarm reset, he asked the firefighters who had pulled the alarm and was told that it was a boy down the hall. The marshal went down the corridor and came face to face with an eleven-year-old boy who investigators had already interviewed as a resident of the apartment building. Unbeknownst to the marshal, families displaced by the fire had been temporarily sheltered in the hotel. The marshal asked the boy about the alarm and was told that the boy had bumped it while walking past. The boy was then asked if his family was on vacation, since he was staying in a hotel. The boy responded that his apartment building had burned and that he and his parents were staying in the hotel for a while. When asked how the apartments could have burned, the boy told the marshal that someone had set a fire on the stairs in two places and then gave a hypothetical description as to how it could have been done—a description that matched physical evidence found at the scene. The marshal calmly bade the boy good night and then called police investigators to report what had happened. The boy was reinterviewed and taken into custody the following morning. Had specific details about multiple sets and mechanism of cause been made public, it's obvious that the boy's remarks to the marshal would have had far less impact.

Whatever the decision about releasing information is, it's important to stick to confirmed, verifiable facts and not speculate about anything. Regardless of whether there may be criminal or civil litigation in the wake of an incident, it's important not to get drawn into the trap of speculating about what may have occurred, especially if the situation remains under investigation.

Remember also that fire departments don't always have lead responsibility for investigating incidents to which they respond. For example, the rail accident that begins this chapter will certainly involve the fire service in firefighting, treating injured persons, and dealing with any release of hazardous materials, but the investigation will be done by rail and transportation officials. A tanker that crashes on the interstate and spills its toxic load will bring firefighters in response but be investigated by the state police. Plane crashes require fire/EMS response but get investigated by federal authorities. It's important for PIOs to recognize this and confine the information they release to their department's area of responsibility and expertise. On the other hand, if a dozen witnesses all say that the plane's left engine was on fire as it went down, there's probably no harm in sharing that with reporters anxious for information, but be sure to attribute such statements to witnesses. There's a marked difference between saying "the plane's left engine was on fire" and "several people who saw the crash said the plane's left engine was on fire."

There will always be questions that you cannot answer. The best way to handle them is by saying "I don't know" or explaining why you can't answer them. If you don't know the answer, try to find out and get back to the reporter before his deadline. If you can't answer the question because it's outside policy or it concerns information that cannot be disclosed, explain why in general terms. Never say "no comment." Those two words have an unbelievably negative impact, leading the hearer to believe that you're guilty or hiding something.

There are many ways to say "no comment" without saying those two damaging words. Consider the following:

"At this point, our investigation is just beginning. Nothing is suspected and nothing has been ruled out. Anything I say would be speculative."

"As you know, we do not identify suspects in arson cases. If there are any developments that we can report, such as arrests, we will notify you as quickly as possible."

"I wish I could answer your question, but we have everything to lose and nothing to gain by releasing further details of the investigation at this point."

It's important to acknowledge that reporters have a right to ask these questions, but decline to answer them as tactfully as possible. The best policy is to answer every question as squarely as possible and to have sound, understandable reasons for not answering those you can't.

Finally, one of the oldest rules of this business is that the names of deceased or seriously injured persons shall not be released until the next of kin have been notified. It's important to know how this gets done in your jurisdiction and to have a working relationship with police and hospitals to know when it's been accomplished in any given case. There will still be some cases that present challenges. For example, if a midday house fire occurs at 123 Ash Street and it proves to be a fatal, odds are that someone else living at 123 Ash is going to hear about it on the radio. The alternative of confirming a fire in the 100 block of Ash is going to panic everyone else living on that block who is away from home.

In some cases, it's best to provide some information before the next of kin are notified. I once responded to a fatal fire in a fraternity house. Once the story hit the wire, it had the potential of frightening every parent in the region who had a son living at the house. As it turned out, the victim of the fire was a female visitor and, by releasing the victim's gender, a general panic was averted.

Similarly, although one wouldn't release the names of those seriously injured in a school bus accident, by quickly obtaining and releasing information as to what school the bus was from and what route it was on, the PIO can avoid alarming large numbers of people who don't have children on the involved bus.

Finally, avoid talking in the fire/EMS dialect. Speak English. This is important for everyone to remember, since good PIOs often attempt to get firefighters to tell their own stories. The PIO may be the one who'll talk to reporters when no one else wants to or has time, but it's best to get the responders themselves on camera or tape, if only to lend a dash of color to the story. Consider the following two sound bites and decide which one says more to the folks at home:

"We were first due, and we initiated a quick attack with an inch-and-a-half preconnect. Using a fog pattern, we made entry into the front hallway."

"We were the first ones here and had to fight every inch of our way into the front hall. With all the heat, it was like trying to go down a chimney."

With all of their other responsibilities, no one expects firefighters to think up sound bites at emergency scenes. Still, all of us should remember that most of the people who will read and hear our words aren't firefighters. To help them understand what we do, we need to explain our experiences in terms that they can easily understand. A little drama and color never hurt, either.

All of the areas covered so far in this chapter apply equally well to both small and large incidents. Large incidents, however, may require a number of additional considerations. Let's say, for example, that a major rail accident involving a release of hazardous materials occurs in your community. As PIO, you may respond to the scene to manage and answer questions from the media, but who's answering the phones back at headquarters? In a major incident, it won't take long for newsrooms hundreds of miles away to look up your number and call for information. These calls may come from radio stations looking for over-the-phone interviews, network or local TV stations trying to decide whether to send crews, and regional offices of wire services such as AP and UPI. The wire services can be a critical means of getting information out to a wide area. Let's say that your community is on a major interstate highway, and a transportation accident involving hazardous materials has resulted in the closure of the road. By getting information on the wire, you can get the word to radio stations around the region that the interstate is closed, allowing them to notify travelers to avoid the area, thereby decreasing your traffic-related problems.

It's important in major incidents to plan on having at least two places for the release of information: the scene and headquarters. It's equally important to maintain communication between the two release points to ensure that both are putting out the same information. For example, if the on-scene PIO is informing reporters that there are twelve dead in a plane crash and the spokesperson at headquarters is releasing one-hour-old information that confirms eight fatalities, there will be confusion and subsequent credibility problems.

Most departments devote few resources to public information, and systems that work well for structure fires and five-car pileups quickly get overloaded when disasters such as plane crashes, haz mat releases, and earthquakes occur. It's important for every department to have a disaster plan and to make public information a part of that plan.

In planning, it's critical to anticipate some of the challenges you may face. In an earthquake or weather-related emergency, for example, there may be no one place where you can anticipate meeting the media. The same may be true of a wildland fire that involves a developed interface area. In such a case, you may have to establish a briefing area near headquarters and gather information from the field for release to the media. It may also be necessary to set up a system to escort photographers in and out of sealed-off areas.

You should anticipate the arrival of large numbers of reporters, photographers, and support staff. The city hall conference room you normally use for news conferences may prove too small, requiring you to bargain for the use of a high school auditorium.

One of the problems that most PIOs experience in major incidents is the arrival of out-of-town reporters. While you may have developed a good working relationship with local reporters, you need to understand that out-of-town news crews don't know you and aren't concerned with long-term working relationships. Predictably, they'll be a bit more mercenary in their newsgathering efforts. In fairness, many network and big-city reporters can be very professional and great to work with, but, generally speaking, the out-of-towners are basically there to get the story at all costs and move on to the next event. It's especially important that PIOs not get so starstruck with the notion that ABC or CNN is in town that they slight their own local media by showing favoritism toward the visiting news crews. After all, the local media is who you'll have to work with when the circus moves on to the next town.

You may have to operate without power or telephones. Enterprising PIOs in major disasters have linked up with amateur radio operators to broadcast information beyond the affected areas so that it can be shared with outside media.

Expect to have to prioritize the information you release. On a house fire, you have the luxury of telling the whole story, including human interest angles such as the name of the victim's pet dog. In major emergencies, you have to set priorities. Your first priority should be the information that people need to know to survive. This information includes instructions on what to do and what not to do, traffic advisories, areas to avoid, what's happening with the schools (if they're operating at the time of the emergency), evacuation and shelter information, how to shut off utilities, and the like.

The second priority is information about the status of the incident—i.e., how many deaths and injuries, how many persons have been evacuated, descriptions of property damage, and what types of actions are being taken by responders in dealing with the event.

The third priority would include human interest stories, acts of heroism, and historical information on events of this nature.

Major incidents typically require the establishment of an Emergency Operations Center (EOC), in which public information officers play a key role. It's critical that major incidents receive public information staffing in the field as well as at the EOC. Most importantly, major incidents require a multiagency response, and fire service PIOs must be prepared to coordinate their efforts with their counterparts in law enforcement and other emergency services to ensure that information is released in a consonant manner. A concept advocated by the Federal Emergency Management Agency is the establishment of a Joint Information Center (JIC), in which PIOs from different involved agencies work together to share and release information. It's critical to maintain public confidence during major emergencies. This confidence can quickly be lost if conflicting information is released regarding the status of the incident or if citizens are given differing instructions as to what to do. If all involved agencies share and coordinate their public information activities in a common center, public confidence can more easily be maintained.

A model emergency public information plan appears in Appendix II. Fire service PIOs should note its basic provisions and ensure that they are covered in local emergency plans.

Chapter Seven

Q and A:
Basic Skills for
News Interviews

The TV studio is hot, and the chief can feel the beads of sweat forming on his brow. It's been two minutes since a technician ran a skinny cord up under his suit jacket and clipped the tiny microphone to his tie.

The request had seemed simple enough: Just do a few minutes live on the morning show to talk about the building of a new fire station and the closure of an old one.

Now, though, he wonders if agreeing to go on was a mistake. His stomach churns and he is sweating buckets. He'd been less nervous the night they almost lost the Crenshaw Building.

The segment's host, an attractive woman about 25 years younger than he, arrives and takes a seat on the other side of the plastic philodendron that dominates the set. She rushes a little as she ruffles through some papers on a clipboard, but she is much cooler looking than he feels. She smiles as she looks over at him.

"This will be a two-minute segment, Chief. I'm just going to ask you a few questions about the new fire station, okay?"

The chief thinks he says "okay," but his voice cracks and he isn't sure what comes out. His mouth and lips are bone dry.

A technician gives a hand signal.

"Quiet on the set, we're on in thirty seconds!"

Interviews take many forms, but learning how to do them is an essential skill in media relations. Whether in a radio or television studio, over the phone, in a coffee shop with a note-taking reporter, or in front of a burning building, interviews are critical opportunities for getting key messages to your community.

An interview is a transaction in which a journalist and a source both come to the table with certain goals. The reporter is looking for information and maybe even has some preconceived notions about the subject at hand. The source is trying to get a specific message across. An interview is often much more than just a conversation. In the most challenging of circumstances, it can more closely resemble a chess game in which both the reporter and the source struggle to make it go the direction each wants it to go.

The vast majority of interviews for fire officials are simple and straightforward—usually a response to questions about an emergency incident or a means to educate the public on some aspect of fire and life safety. Accordingly, it's important not to get so psyched out that you view every interview as a potential *60 Minutes* encounter. There's a world of difference between describing the

importance of regularly testing smoke detectors and serving as a department spokesperson after a engine company
is involved in a fatal accident while responding to a call.

Regardless of whether an interview comes about
because you've initiated a story or the media's come to you
with a request, the basic process is again much like fight-
ing a fire.

Begin with a size-up. Who's asking for the interview and
who is the audience? Is it a raucous phone-in radio show
with an abrasive host? Is it one of those staid, lengthy
Sunday morning public affairs programs? What medium is
involved, and how long will the segment last? Will it be
live or taped?

Do your homework. Know the topic the interview's
going to be about, and ask yourself whether you're the best
person to speak on that topic. A successful interview is a
combination of good presentation skills and knowing the
subject. If, for example, a magazine for architects wants to
do a technical story on recent fire code amendments, it
may be better for a fire marshal to do the story than a PIO
who has limited knowledge in that area.

An interview is an opportunity to reach a large number
of people, and you should go into it with a specific objective. If the interview has been prompted by an incident,
your objective may be to get across basic information
about what happened and what's currently being done.
On the other hand, if you're there to promote smoke
detectors or a capital improvement program to upgrade
your department's facilities, you should go in with two or
three key points you want to get across.

The key steps in preparing for any interview include:

- Have a sense of what's newsworthy about your topic.
- Know what you want listeners, readers, and viewers to
  do in response.
- Decide what two or three key points you want to make.
- Plan how you'll support your key points, such as with
  statistics or examples.

- Anticipate what questions you'll probably be asked, and think about how to steer your responses back to your key points.
- If you are nervous about the interview, do practice interviews with coworkers.

There's a balance to be found between simply responding to an interviewer's questions and being so forceful in making your points that you come across as aggressive or pushy. One technique recommended by interview trainers is that of bridging, in which you acknowledge a reporter's question and then bridge into a point that you want to get across. If you think about it, you've probably heard this done many times on radio or television.

Reporter: "Isn't there public concern about the cost of buying a ladder truck?"

Chief: "Cost is certainly a factor, but (bridge) we're more concerned about the number of people we now have living and working in high-rise buildings that we can't reach in case of fire."

As discussed in Chapter Two, print is the medium that carries the most information. Accordingly, newspaper reporters ask more questions and seek more detail than their broadcast counterparts. The good news is that print interviews are usually the least intimidating in terms of hardware and paraphernalia. A typical interview for publication will entail a one-on-one conversation with a journalist taking notes or possibly using a tape recorder. The interview may be over the telephone, in which case you should bear in mind that many states allow unacknowledged tape recording of phone conversations as long as one of the parties knows it's being recorded. Be prepared for print interviews with lots of detail, and provide reporters with copies of relevant documents.

Radio interviews can be a little more intimidating, since radio isn't just a matter of what information you provide but also how you sound. We've all heard bad interviews in which the subjects "uumm" and "aaah" their way through their message. The people who come across well on radio are those who speak clearly and strongly and simply state their case. Avoid speaking in a monotone. Project and express yourself.

A radio interview can take place in the field, over the phone, or in a studio. In each case, the basic principles are the same. Unless you're doing a Sunday morning public affairs program, you should make your points as succinctly as possible. Radio news relies heavily on sound bites: short, pithy quotes that explain things in about fifteen seconds or less. Summarize and lead with your key points. Keep it simple. When talking to a radio reporter over the phone, close the office door and turn down the scanner. Focus on the interview and don't get distracted. I find it useful to stare at something on the wall or close my eyes.

Television is the most intimidating medium. A TV news crew comes to your office to interview you and, all of a sudden, it isn't your office anymore. Now there are bright lights, cables and cords all over the floor, a camera on a tripod, and a microphone shoved in your face. Your mouth gets dry, and your forehead gets wet. Television's not just about information and how you sound, it's also about how you look.

Remembering a few small pointers can help you get through a television interview. Hard as it may be, ignore all the equipment and focus on the reporter. Imagine yourself in a one-on-one conversation. Look the reporter in the eye. Speak conversationally and naturally—you'll come across much better to the viewers at home.

Because television is a visual medium, you should give some thought as to how you look. Unless you're being interviewed on your fourth day of dealing with a wildfire or some other disaster, you'll want to be well-groomed and clean-shaven. Don't wear sunglasses or photo-grey prescription glasses, since they'll make you look shady and suspicious.

It isn't necessary to wear a full dress uniform or turnouts

for a TV interview, but you should wear something with your agency's name and logo on it, whether it be a baseball cap, polo shirt, sweatshirt, or jacket. It's a simple form of marketing your agency and, if you think it's corny, consider that a number of TV stations around the country are now issuing their field reporters baseball caps and windbreakers with station logos to wear on camera. If you're in a situation where you're on call or often wear civilian clothes to work, it can be useful to keep a few such items in your car so you'll always be ready for when the TV crews arrive.

If you're wearing civilian clothes, it's important to look nice, but you don't necessarily have to be dressed up. If several hundred people have been evacuated in the face of a major wildfire, viewers will have trouble finding credibility in anyone who shows up at the scene in a three-piece suit. When I was a nonuniformed police PIO, I did a lot of interviews with my sleeves rolled up and my tie loosened, giving the subtle (and honest) impression that I was working. In any interview situation, people relate better to spokespersons who come across as real people projecting warmth, empathy, and appropriate humor rather than the stiffness and nondimensionality of bureaucrats.

If you're doing a formal, in-studio TV interview, remembering a few clothing tips can be important as well. Solids are better than stripes or other busy, distracting patterns. Neckties should be simple in design. Jewelry should be kept to a minimum, as should anything else that may detract from you and your message. If you're going to cross your legs, make sure your socks are long enough to avoid a gap between their tops and your pant legs. A summer-weight suit will be more comfortable in the heat of studio lighting than a heavy one.

Finally, consider a piece of advice that I received early in my career: Never look up and never look down on camera. If you look down, it appears as if you're hanging your head in shame. If you look up, it appears as if you're imploring God to give you the answer to the question.

Beyond questions of style, a number of rules pertaining to content apply to any interview:

1. Honesty is the best policy. Your credibility is the most important thing you have with the media. You must always be truthful, and in situations where you can't tell the whole story, the sum of what you do say mustn't be misleading.

2. Speed is second only to accuracy. The most articulate spokesperson or the best supporting arguments are useless after deadline.

3. Always assume you're on the record. *On the record* means that whatever you say can be printed or broadcast and attributed to you. It is the safest condition under which to do an interview. In the major leagues of media relations, there are some other avenues by which information gets released, but you should exploit them only with the greatest of caution. *Background* means that a reporter can use the information but must attribute it to a nonspecific source, such as "a fire official." *Not for attribution* means that the reporter may use the information but cannot attribute it to any source. If remarks are made *off the record*, it means that the reporter may not use any of the information he hears. This is usually done in cases where it's necessary to help a reporter grasp the full meaning of a complicated situation that can't yet be made public. The problem with this method is that enterprising reporters will sometimes use what they get in this manner to leverage the same information from another source on the record. Going off the record, someone once explained, is like other things in life: You should only do it with someone you know and trust, and never on the first date. In any case, if the rules are anything besides on the record, *they must be agreed on by the reporter and the source before the interview begins.* Don't expect to

give away a major secret during an interview and then retroactively declare it to be off the record.

4. Don't assume any knowledge on the reporter's part. Unless the reporter is someone you know to be experienced in covering your agency, don't assume he knows anything about your business. Take the time to explain the issue and make sure he understands, especially if he asks questions that stray from the subject.

5. Stick to confirmed, verifiable facts. I call this "doing the play-by-play and avoiding the color commentary." This is especially true in situations that remain unresolved or under investigation. Avoid speculation, and simply pass along confirmed facts. "At 1:32 a.m., firefighters responded to a reported fire at 123 Elm Street. On arrival, they observed a large body of fire on the second floor of a two-story house. Firefighters entered the building to attack the fire and search for victims. They found a victim in the house who was pronounced dead at the scene, and they controlled the fire in seventeen minutes. Investigators are now on the scene."

6. Deal only with your area of involvement and expertise. Suppose that a tank truck is involved in an accident and dumps a hundred gallons of methylethylbadstuff into a lake. A logical question for a reporter to ask might be what long-term environmental impact the spill will have on the lake's ecosystem. Unless you hold dual degrees in chemistry and wildlife biology, your safest course is to summarize whatever hazards your guidebooks tell you the chemical has and explain what's being done to contain the spill.

7. If you can't comment on something, explain why. As mentioned above, two of the most damaging words you can ever use are "No comment." It's better to explain that "investigators have just arrived

on the scene to determine how the fire started," or "the information you've asked for is medical in nature and the law prohibits me from releasing it," or "we have an ongoing investigation and any release of information might compromise its integrity."

8. It's okay to say "I don't know." It's part of being honest. But you get bonus points with the media if you can direct them to a source who does know or if you otherwise find out the answer before deadline.

9. Speak simply and avoid jargon. A child in middle school should be able to understand you. Don't talk about "a medical branch that was established to perform triage"; instead, talk about "paramedics who were assigned to see how many patients we had and how badly they were hurt."

10. Be as open and cooperative as possible. The media and the public respect this trait.

11. Project calmness and control. Given what we encounter in this business and the pressure of responding to the media's demands, this can be difficult, but the last thing the viewers at home need to see during a disaster is a stressed-out public safety official who appears to be losing it.

What about trick questions? Although most interviews will be simple and straightforward, and the majority of reporters aren't out to get you, there may come a time when you'll face a question designed to trap you in some way. Some of the more common traps are as follows:

1. The intentionally rude question. Some reporters act abrasively in an effort to get you to respond in kind. Remember, they can edit themselves out of the tape—you can't. If you lose your temper, you'll lose whatever control you have over the interview. Stay calm and polite, address the reporter by his first name, and smile if it's appropriate.

2. The leading question. "The fire at ACME Chemical shows how inadequate your inspection program is. How can the public have confidence in your prevention activities?" If you know your business, you should be able to answer along the lines of "Well, we conducted more than 1,500 business inspections last year and found some 1,200 violations. I think our fire record would be much worse if we didn't do the kind of aggressive code enforcement we do."

3. Responding to information you haven't heard firsthand. "A spokesman for the Homebuilders Association charged earlier today that your proposed residential sprinkler ordinance will price a lot of people out of owning a home." If you haven't heard or read the charges firsthand, don't take the bait. "I'm afraid I didn't hear those comments. What I do know is that, for less than the price of wall-to-wall carpeting, residential sprinklers are the best way to make people fire-safe where they live and sleep."

4. The pause after your answer. Some reporters have mastered the technique of not responding after you've answered their question. Like Nature, many of us abhor a vacuum and will continue talking, perhaps saying things that dilute our message. Answer the question and, when you're done, stop talking. You'll never hear the dead air on the news.

5. The accusatory question. "There have been seventeen arson fires in this neighborhood since the first of the year. When is the department going to do something about it?" Stay cool. "Since the pattern emerged in mid-January, a team of police and fire investigators has spent hundreds of hours working on these fires. It would compromise their work to talk specifically about their activities, but people need to know that a lot is being done."

6. Unacceptable alternatives. "Are you in favor of new restrictive fire codes, or do you want a com-

munity so open to new business and development that large fire losses are inevitable?" Don't accept either alternative. "What we have to do is find a balance that makes our city fire-safe and attractive to new businesses."

7. A reporter's negative statements. Rule One: Never repeat them. If a reporter says, "People tell me that your department's paramedics are the most poorly trained in the state," don't respond by saying they're not the most poorly trained in the state. Say instead, "To be certified as paramedics by the state, our personnel receive so many hours of training a year. This training is mandated and audited by the state to ensure that quality care is delivered on the street."

It's rare that you'll encounter these kinds of trick questions, but if you do, it is most important to stay calm and cool. Be positive, not defensive. If you don't let a reporter put words in your mouth, you won't appear to agree with something you don't. And don't be argumentative, lest the emotion of your response become more newsworthy than the issue under discussion.

Above all, the best training for doing interviews is to do interviews. Take every opportunity you have to interview with reporters. You'll find that, with time and experience, your skills will improve and your nervousness will diminish. The experience you'll gain in making your key points simply and succinctly over noncontroversial matters will pay off when more challenging circumstances eventually arise.

Chapter Eight

When the News Hits the Fan: Preparing for the Inevitable

It was nearly closing time, and the bartender cast a wary eye at the group in the back booth. They had come in around eight o'clock and had been drinking heavily ever since. The group was made up of large, athletic-looking men, several of whom had grown more belligerent over the past hour. The whole group had become loud to the point where other customers had moved away from them.

As the bartender looked up at the clock and idly began wiping a glass, he suddenly heard shouting and the sounds of breaking glass and furniture from the back of the bar. A brawl had broken out in the back booth.

A police dispatcher answered the bartender's call for help on the second ring. When asked if he could describe the men involved in the fight, the bartender said, "Most of 'em are all wearing the same T-shirts. They're navy blue with some kind of cross design on the left chest and Metro F.D. on the back."

Bad things do happen to good departments, if for no other reason than that they're staffed by fallible human beings. Citizens complain, equipment breaks down, controversies arise, normally good employees do dumb things, bad tactical decisions get made, and sometimes the ball just bounces the wrong way. Like people, even the best of organizations has an occasional bad day.

At the same time, the best way to avoid bad news is to prevent bad things from happening. Those concerned about a fire department's relations with its customers and constituents need to be constantly vigilant about the organization's training, discipline, and all other forms of institutional maintenance.

How much impact bad events have on fire departments has a lot to do with how they're handled when they occur. The most realistic attitude a fire service manager can take is to accept the notion that bad news will happen and plan in advance for how to handle it.

Advance planning has several key components: maintaining positive media relations, identifying potential problem areas, and knowing the rules ahead of time. Let's look at each of these areas.

Your relationship with the media is critical before, during, and after a crisis. In advance of a crisis, a reputation for being candid, helpful, and honest can give you an edge that will stay with you when things go sideways. One of the things ongoing public relations efforts do is build a reservoir of goodwill against the inevitable controversy or departmental misstep. It doesn't mean that the media won't cover the story and ask hard questions, but it does mean that you'll certainly be given a chance to tell your side and, in grey areas, you may get some benefit of the doubt. It's much easier to deal with bad news when you've established credibility and are on a first-name basis with the people writing the stories.

On the other hand, if your department is one of those that regards reporters as annoying busybodies and rarely returns their calls, the coverage may be less than sympathetic. In fact, the first you know of the problem may be when it appears on page one, with six paragraphs' worth of quotes from an outraged citizen and a concluding line to the effect that "city fire officials did not return calls Tuesday afternoon."

It's also important to maintain a positive relationship during the crisis. All reporters have horror stories about public officials who are readily available when the news is good and disappear when things go bad. Nothing can test you more than trying to be helpful and polite to people who are reporting information that will publicly embarrass your department, but it's important to maintain your composure and poise. If you lose your cool, you lose whatever control you had of the story. We've all seen people get nervous, defensive, or openly hostile on camera, and we've all come away with negative impressions. Remember the adage "Don't argue with people who buy ink by the barrel or videotape by the case."

At the same time, if you think coverage of the story is truly unfair (a judgment that you should make only after thoroughly and objectively weighing your own obvious biases), there are a few things you can do. First, you need to analyze whether the story contains errors in fact or whether you're upset about the tone. Factual errors are fairly simple to correct, especially if you have documents that show that the story varies from the truth. Story tone is much more subjective. Veteran reporters will tell you it's not uncommon to do stories about controversial issues and have advocates for both sides complain about the coverage. Ironically, this can sometimes mean that the story was pretty well balanced.

In either case, if you want to follow up, your first step is to talk directly to the reporter who produced the story and calmly explain what your concerns are. Reporters are no different from the rest of us, and if we're unhappy with something they've done, they'd rather we talk to them directly than go straight to their boss.

If you and the reporter fail to come to some kind of understanding, and if you think it's important enough to pursue further, write to the editor or news director, outlining your concern and summarizing your contact with the reporter. Do not do this in anger, and be as objective as possible in composing your complaint.

You also need to keep some perspective. The story will eventually die, and you'll still have to live with the media. Don't burn bridges, and don't give in to the temptation of taking shortcuts with the truth or any other facet of media relations. There will be other days and other stories on which you and the media will have to work together. Like a marriage, your relationship with the media will have its ups and downs. There will be squabbles and times when one party will really upset the other, but the focus should be on maintaining the long-term relationship.

It's also important to maintain relationships after the crisis. Media relations can be sabotaged in many subtle ways. Let's say Channel 7 does an embarrassing story on the off-duty altercation described at the beginning of this chapter. All of a sudden, dispatchers are always too busy when someone from Channel 7 calls. Department members refuse to speak with the station's reporters.

Given time, the friction increases. Every chance the department gets to snub Channel 7, it does. Every time Channel 7 gets to take a shot at the department, it does so with barely concealed glee. Don't trade short-term vengeance and satisfaction for long-term losses.

The second phase of planning for a crisis is anticipating what kind of bad news could come out of your department. For most departments, this includes losing buildings and, in the worst case, people to fire. It also includes personnel problems, mechanical breakdowns, code enforcement controversies, poor response times, and emergencies for which the department wasn't prepared. You need to think in advance about how you might respond to each of these situations. For example, if a piece of apparatus were to break down at a critical time, would you be able to quickly develop statistics showing how rare such breakdowns are or detailing the number of miles or hours of operation that have passed without incident? Would you be able to gather and summarize information on your department's apparatus maintenance program and explain, briefly and in nontechnical terms, what your organization does to prevent equipment breakdowns?

Get in the habit of looking at other organizations in the media hot seat and ask yourself how you'd respond under the same circumstances. If those organizations are fire departments, you may well have to respond to local "could it happen here?" inquiries.

Fire departments must know the rules before bad news or a crisis hits. Policies must be in place that spell out who speaks for the department and what information will and will not be disclosed. For example, in the hypothetical situation that begins this chapter, it is possible that disciplinary action will be taken against the firefighters involved in the off-duty altercation. But can you release specifics about discipline, or does that constitute personnel information that cannot legally be disclosed? It's going to depend on department policy, which should be based on local public records law.

Based on the experience of many organizations, there are a number of considerations involved when your department falls into a bad-news situation:

1. The media will do the story with or without your help. You'll probably like the results better if you get to present your side.
2. If it's obvious that a bad-news story is going to get out, an organization is usually better off if it breaks the news before reporters do. You'll be in a much more defensive posture if you've sat on a scandal for three days before the *Daily Bugle* comes around to confirm bad rumors they've heard elsewhere. It's hard to give yourself a black eye, but if you're going to get hit anyway, you can be in a better position if you do it yourself.
3. Have someone do fact-finding to determine what information has been confirmed and what is just gossip and rumor.
4. If you stall, stonewall, lie, or attempt to cover up a situation, you only make the situation worse. It gets back to what Mom taught you: Tell the truth and confess. The citizens may think you're inept, but at least they'll still trust you.
5. The best way to end a bad story is to disclose as much as possible as quickly as possible. The natural temptation is to disclose as little as possible and hope that the rest of the bad news doesn't get out. That never works. If the story is one that has attracted a lot of media attention, you'll just find yourself in front of the TV night after night as anchors breathlessly report, "Eyewitness News has learned that..." and "Contrary to what fire officials said Wednesday...."
6. If the situation involves lawyers, work closely with your own. Attorneys need to understand that it's possible to win a case in court but still lose in the arena of public opinion. Often the first advice an

attorney gives a client being sued is not to say anything. The problem is that the civil complaint filed in a lawsuit is a matter of public record, so what appears in the paper the day after it's filed is paragraph after paragraph of allegation and accusation, with a concluding line that "fire officials refused to comment on the matter." There are two problems with this: It makes your public think you're guilty, and it gives the other side a free throw.

7. Do express concern when bad things happen, and explain in general terms what's being done about it. You can be concerned about a situation without having to determine whether anybody is guilty or innocent ("The department is extremely concerned about the allegations of sexual harassment at Engine Company 23, since we strive to maintain a harassment-free workplace") and, at the same time, explain what's being done ("Chief Smith has appointed a five-member board of inquiry that will investigate these allegations and report back within two weeks"). Outline what further actions will be taken ("The results of this inquiry will be released to the media after the report has been presented to the chief, and disciplinary action will be taken if it appears appropriate") and explain why no additional comment will be made at this time ("What we have at this point are unsubstantiated allegations, and it wouldn't be appropriate to comment on them further until such time as they can be verified and investigated"). If appropriate, it doesn't hurt to review actions taken to prevent something bad from happening ("Over the past year, all members of the department have participated in mandatory training exercises designed to educate them on the issue of harassment and help them understand how to prevent it").

8. Don't let yourself get drawn into a negative "ain't it awful" reaction. If the situation allows, focus on

what steps have been taken to prevent an occur-
rence from happening again ("Effective immediately, our procedures are being changed, and no
department vehicle will be left unattended with the
keys in the ignition").

9. Have background materials prepared on the different parts of the department, including their history,
the work they do, and how they're staffed. If, for
example, a situation arises involving the hazardous
materials team, it can be useful to give reporters
information about when and why the team was
formed, how many members it has, the kinds of
training they receive, what equipment they carry,
and how many calls they responded to last year.

10. Don't overlook internal communication. It's easy to
get distracted by inquiries from reporters, elected
officials, and citizens, but don't forget to ensure that
the department's members are kept informed about
what's going on. They have as much right to know
as anyone else and are often key communicators on
and off the job.

Finally, stick to the basics. Be accessible to reporters, tell
them what you can, and have sound reasons for whatever
you have to hold back. There's a lot to be said for telling
the truth and doing the right thing.

Chapter Nine

Connecting With the Community

The captain stood in the doorway of the day room. It was seven o'clock, Tuesday.

"We've got the Eastside Neighbors meeting here in half an hour. Bill, you and Dave get the community room cleaned up. Mike, how about throwing together a couple pots of coffee?"

In the late 1980s, a revolution began sweeping through American law enforcement, perhaps the first one to carry a profession backward and not forward. At the start of the century, police had patrolled neighborhoods on foot, knowing intimately the communities and people they served. Over time, in the interest of efficiency and cost containment, a more reactive model of policing developed. Cars could patrol larger beats than foot officers could. Placing radios in those cars enlarged patrol districts even more. The advent of 911 made it easier to report incidents; thus, the volume of calls increased. Policing became a matter of responding from one incident to the next, and citizens rarely came in direct contact with the police except in emotionally tense settings, after something had already gone wrong.

A newer concept, known as community policing, acknowledges the rift that a reactive, incident-driven, service-delivery model has created between police officers and those they serve, and it suggests strategies for change. While described in different ways by academics and law enforcement officials, there are several key concepts inherent in almost all community policing programs.

One of these is problem solving. Rather than simply responding to the same calls in the same neighborhoods over and over again, officers are expected to develop longer-term solutions to neighborhood problems, ideally in partnership with area residents, businesses, social service agencies, and other institutions.

Another basic tenet of community policing is to increase the number of nonincident-related contacts that officers have with citizens. By allowing officers to meet citizens at neighborhood meetings, home security surveys, and community events, the officers and citizens have a better opportunity to get to know each other as real people and better understand their respective roles in resolving problems.

Key to the philosophy of community policing is the policy that officers work with, not just for, the community. Partnerships, participation, and community involvement are all critical to success.

Why so much about law enforcement in a book for the fire service? Because many of the lessons being learned in community policing are applicable to our environment. As a fire officer placed by circumstance into a police PIO's role, I had the opportunity to see community policing initiatives being introduced into a midsized city. Some of the initiatives were largely a matter of style, such as logos and slogans. Others were of considerable substance, such as bicycle patrols and storefront police stations. Not all of them were universally accepted by police officers, and much of the work was based on trial and error, but the community responded positively to the changes. While community policing has been practiced in different communities with varying degrees of success, it grew out of many of the same concerns that we in the fire service have about our relationships with those we serve.

Much of what we do is reactive in nature, and most of our contacts with citizens involve either emergencies or enforcement activities, such as fire inspections. A small percentage of those who pay our bills have regular face-to-face contact with us.

The people we serve have a limited understanding of what we do. Like the police, most people form their opinions of us based on what they see, read, and hear in the news or entertainment media.

Although our issues are different, we operate in the same community and political environments as law enforcement. The past decade has seen significant change in the political environment, and fire service managers place themselves at a disadvantage if they overlook some of the trends apparent in many communities.

Political activity is intensifying, albeit in smaller geographic areas. Tip O'Neill once observed that "all politics is local," and this has become even more so in growing communities. As cities grow, it becomes harder for the average citizen to relate to them as a whole. People don't live in Manhattan anymore; they live in Soho, Greenwich Village, or on the Upper West Side. The growing belief that an average person can't make a difference in a large community has been balanced and replaced with renewed neighborhood-level activism. The old-money patrician politicians with communitywide vision are being replaced by an increasing number of neighborhood-specific activists and single-issue candidates. What this means to the fire service is that a citywide fire station relocation study won't draw nearly as much attention as the proposed closure of a particular neighborhood station.

Many people report a growing lack of confidence in government, whether or not it's deserved. Long accustomed to being the good guys, the fire service has often been surprised when it's come under the same sort of criticism historically reserved for "politicians" and "bureaucrats." The roots of this phenomenon may well be in national scandals such as Watergate and the increasing tendency of the news media to do investigative reporting, but some of the responsibility can be traced to the political process itself.

Negative campaigning has become a major part of the American political scene. Many candidates for office run attack ads against their competitors. Studies show that negative political campaigning works in many races. Unfortunately, when challengers run negative attack ads against the very offices they're running for, a documented side effect is decreased confidence in government.

There's also a paradox here, related to the localization of politics described above. In surveys, people will give low ratings to institutions but high ratings to the local embodiment of those institutions. For example, they may have little confidence in the state legislature, but will give high marks to the representative from their district. They may have their doubts about the school district and its board of directors, yet still like the neighborhood elementary school and its staff.

In many cases, the source of these sentiments about government stems from perceptions of federal and state government, but the greatest impact is at the local level, where most fire departments politically exist. Unlike law enforcement, there are no state fire departments in the United States. People have fire protection because they've acted locally to provide for it. Americans don't get to vote on federal or state budgets. Most of us will never testify before a U.S. Senate subcommittee, and the governor probably doesn't pay much attention to our letters to the editor, but it's a very different game at the local level. Local government is by nature more accessible to those it serves and more dependent for its continued existence on voter goodwill and approval of budgets, levies, and bonds. At the local level, we in the uniformed services also have the advantage of being positioned to make regular, actual differences in the lives of those we serve.

Other trends in local politics include changes in family demographics and the pace of life. The increase in single-parent households and two-parent households in which both parents work, coupled with the increasingly frenetic pace of life, have reduced the capacity of many Americans to serve as volunteers, hold elected office, or otherwise play active roles in community affairs.

Local government has also been increasingly fenced in

by mandates and voter-imposed restrictions. While nobly intended and laudable, the dictates of state and federal government in such areas as occupational safety and health, hazardous materials right-to-know legislation, access for the disabled, and fair-labor standards have all cost fire departments time and money. The fire service has imposed its own mandates as well. Standards are constantly being created by national fire service organizations, such as the NFPA, and other bodies, such as state accreditation programs. Even when these standards are voluntary, they create liabilities for fire departments. It doesn't take much imagination to picture an attorney asking a fire chief on a witness stand if he's familiar with a model standard on firefighter safety. At the same time that mandates and standards are complicating our profession, voters in many parts of the nation are imposing property tax limitation measures that reduce the capacity of local governments to meet these mandates and still perform their primary work.

This situation hasn't stopped citizens from asking more of government, even as resources have diminished. "Doing more with less" has become a familiar cliche to most fire chiefs. For most fire departments, the past two decades have seen a tremendous expansion of our service portfolios as we've added EMS, hazardous materials, and emergency management to our repertoire of missions.

Racial and ethnic diversity creates additional issues for fire departments wishing to be truly responsive to their communities. The reality of a diversified population creates a need for work, not only in making public information and education multilingual and multicultural, but also for validating recruitment and promotional practices.

A phenomenon related to the news media's growing emphasis on investigative reporting, and one noted by many media critics, is the increasing tendency of journalists to focus on the politics of issues rather than the issues themselves. Much of the election coverage we get tells us more about who's ahead in the polls or how important it is for a candidate to win a particular ward or precinct than what the issues are and what can realistically be done to manage them.

This is further complicated by the continued shortening of our national attention span, as well as the competition of new forms of media for the limited time we have to gather information. Network television has been supplanted by cable, local newspapers compete for our attention with national editions of big-city papers, and media scholars say the Internet will eventually become the most dominant communications medium—one that functions without trained intermediaries such as reporters and editors to verify the accuracy of the information placed on it. This explosion of information has made it possible for someone to consume vast amounts of news on a daily basis, with no guarantee they're accessing any local news.

More than ever before, our messages need to be brief, relevant, and to the point. The annual report may be the chief's pride and joy, but it's probably not getting read cover to cover by anyone outside the fire department anymore.

In the midst of all this change, how can fire departments better connect with their constituents? Should we even think of them as constituents? A growing trend in the fire service has been to identify the people we serve as customers and to focus on ways to improve customer service.

The notion of customer service is, in many ways, a positive influence on the fire service, and it applies well to individual contacts in emergency and nonemergency situations. Still, fire departments must always bear in mind the unique relationship they have with their "customers."

In a lunch-hour transaction between a sidewalk hot dog vendor and a hungry office worker, the office worker exemplifies the word *customer*. Our customers, by contrast, not only collectively own our assets, they also have legally proscribed access to most of our records and meetings, and they occasionally get to vote on the economic future of our business. The schools of government and business at Harvard are housed in different buildings for good reason. Consider the position of community policing theorists, who hold that all citizens are responsible for community order and safety, and that police officers are differentiated only in that they have full-time responsibility for these matters.

Not all citizens, especially those who have had their fill of corporate downsizing and outsourcing, or who view themselves as being more than just consumers, appreciate the business analogies so often used in government. For some people, the term *customer* creates a disenfranchising cynicism when applied to public affairs.

Semantic concerns aside, the initiatives taken by fire departments under the umbrella of customer service are exactly the kinds of activities that increase support for public fire agencies. Increased care for and assistance to people experiencing emergencies, a willingness to handle needs above those met by basic response, and a continual search for ways to enhance service to the community are the hallmarks of the customer service movement. Consider a few examples from my department:

A teenage girl is driving into Portland from a hundred miles away to pick up her younger sister at the airport. On her way to the airport, the big sister is involved in an accident on the freeway. Her car is damaged and she receives minor injuries, thus requiring transport to a hospital. The captain of the responding engine company uses a cell phone to call headquarters and ask for help. Within minutes, personnel at headquarters have called the mother and the airline to inform them of the situation. A staff officer heads for the airport, contacts airline personnel, picks up the young traveller, takes her to the hospital, and stands by with the reunited siblings until the mother arrives.

A man is painting his house when he suffers the signs and symptoms of a heart attack. Responding firefighters and firefighter-paramedics treat the man at the scene and load him into an ambulance for transport. When the ambulance leaves the scene, the firefighters take fuller notice of the project that the man had been working on. The brushes are still wet, the paint cans open, and it will be some time before the man is able to finish what he's started. Before returning to their station, the firefighters finish painting the house.

A devastating fire strikes an apartment building. Long after the fire has been extinguished, firefighters remain on the scene, helping victims remove salvageable items from what's left of their apartments. A nearby firehouse has a vacant bay on its apparatus floor, and firefighters there volunteer the space as a temporary storage facility for tenants until they can find new housing. Fire district cell phones are loaned to the local Red Cross and a church involved in the relief effort. A financial contribution to a fund for fire victims is made from a community assistance trust maintained by the fire district, and the International Association of Fire Fighters local contributes as well.

A field guide prepared by Tualatin Valley Fire and Rescue provides guidance for line personnel on how to help people, beyond the basics of emergency response. The guide includes a series of questions that members can ask themselves to evaluate what sort of assistance can be provided. Are people left stranded, without transportation? Do they need temporary shelter or food? Is there a need for crisis intervention or emotional first aid? Has personal property left at the scene been secured?

The field guide includes a list of district programs available to help people, such as a community assistance trust fund that allows company officers to spend up to one hundred dollars to help someone encountered on a call. In addition, the guide has an extensive listing of local non-profit organizations providing help in such areas as drug, alcohol, and domestic abuse; mental health services; care of injured or deceased animals; youth services; and foreign language resources.

The above are just examples of things that fire departments can do to enhance their service. Many departments around the country are engaged in similar initiatives. Whether such activities are undertaken in the spirit of "serving the customer" or simply doing the right thing, they are actions that can give a fire department a strong reputation in the community and build public support.

Beyond what firefighters themselves can do to assist people, many departments use other people to provide crisis intervention and support for victims. Tualatin Valley Fire and Rescue has a chaplain program in which several local clergymen carry pagers so that they can respond to calls and provide assistance to victims. If a victim or involved family has religious convictions, the chaplains can minister to them or connect them with their own church. If not, the chaplains provide nonsectarian crisis assistance. They also provide crisis counseling, transportation, and next-of-kin notification, and they serve as liaisons with the local Red Cross in situations where fire victims need assistance with shelter and food. Even beyond that, the chaplains not only provide service to victims on behalf of the fire district; by helping families through the immediate aftermath of traumatic events, they also enable responding firefighters to clear the scene and be available for additional calls.

Next door, the Portland Fire Bureau provides similar services in different ways, as one of a number of cities involved in *TIP*, or Trauma Intervention Program. Citizen volunteers receive extensive training in crisis intervention before taking turns being on call to respond at the request of police and fire officials.

Another consideration for fire departments is that all resident taxpayers contribute to their operations whether they use them or not, a further divergence from the customer analogy. Just as senior citizens who don't have children in school sometimes balk at voting in favor of school levies, so too do people who rarely experience fires or medical emergencies question community expenditures on maintaining certain levels of protection. The question "What's in it for me?" is an indigenous, universal aspect of human nature.

It's important to recognize this factor and do things that add value to the emergency response and code enforcement activites that you already provide. This is where public education activities can play a major role. Getting speakers out in the community to talk about fire safety and injury prevention, to teach CPR classes, or to organize neighborhood emergency response teams are all ways that fire departments can demonstrate their value to citizens who don't experience emergencies. The same value is demonstrated when a fire department sets up an information booth at a community festival, in a shopping mall, or at a school event.

A unique initiative taken on several years ago by my department is a child safety car seat program, in which firefighters inspect child safety seats on a drop-in basis at fire stations and at periodic community clinics, checking for proper fit and installation, as well as for recalls. The inspections do take time, but they also increase the number of positive contacts that firefighters have with citizens, plus they add value to the service we provide the community.

Throughout the country, other types of programs have been initiated by fire departments seeking to enhance their service and become more proactive in their communities. These have included smoke detector giveaways, immunization clinics, blood pressure screenings, and programs in which firefighters adopt local schools, serving as positive role models by spending time in classrooms and lunching with students. The Fire Department of New York has a program in which children are trained to view firehouses as safe places to go if they have been injured or threatened, and other fire departments have signed up their stations as part of established Block Home programs.

Another way to get citizens involved with the fire service is to invite them to participate in the development of long-range or strategic plans. As more departments hone their ability to think strategically, they are developing goals and objectives to guide them over extended periods of time. While much of what we do is technical in nature, and public involvement may complicate the planning process, involving citizens in the development of such plans is a way to educate them on the nature of our business and give them a greater sense of connection with and ownership of our agencies. Such involvement can take place at the grass roots neighborhood level, or it can involve the formation of a blue ribbon advisory committee, on which local business and community leaders are asked to serve.

As an aside, it's critical for PIOs to be aware of such strategic goals and objectives so as to highlight them in news coverage, when appropriate. As this book was being drafted, my department was involved in an effort to amend building codes to require fire sprinkler systems in all new multistory apartment buildings. Although local newspapers had given the story some attention, trying to get radio and television to cover a code issue proved difficult. When two simultaneous apartment fires occurred on the same weekend morning, sending three residents to the hospital, we got the attention of radio and television. By highlighting the sprinkler initiative in news releases and interviews about the fires, we got coverage of the issue and raised public interest just days prior to an important hearing on the proposal.

Community visibility is also critical for any fire department. Elsewhere in this book, we talked about making sure that the department's name is boldly marked on apparatus, T-shirts, sweatshirts, and turnouts. It's also important that the department's name and logo appear clearly on all educational materials, inspection forms, permits, and other documents to remind citizens who their fire service provider is.

Visibility takes many forms. Some departments go so far as to encourage or require their officers to join service clubs, such as the Rotary Club, Kiwanis, or Optimists. Other departments join the local Chamber of Commerce. Participation in these organizations creates networking opportunities and builds relationships with different factions of the community. It's a lot easier to approach a local insurance agent for help with a smoke detector program if you're already on a first-name basis at the Lion's Club.

Fire departments also get visibility when they recognize the efforts of others in the community who support their mission. If a citizen helps people evacuate a burning building, performs CPR, or performs first-arrival firefighting with a garden hose, it does wonders to see that he is recognized. Similarly, it never hurts to see that elected officials or corporate partners who help with prevention efforts get plaques or department sweatshirts in recognition of their support. Media coverage of such awards events only enhances their value.

As mentioned earlier in this chapter, the Internet provides an additional source of visibility. The rapid growth in fire-related websites and home pages suggests how powerful the Internet has become in a short time. Fire departments can use websites to share information about their agencies locally and worldwide, to provide safety tips and information, and to disseminate news releases. In major emergencies or seasonal situations, you can forestall repeated public inquiries by posting information on the Internet. Say, for example, that your region is hit by severe winter weather. By posting cold-weather fire prevention and life-safety advice on the Web and publicizing its availability, you can allow people to get lifesaving information at their own convenience. Several fire departments have gone so far as to place their Internet address on the sides of their apparatus. The only downside is that community access to this technology is going to vary from place to place. In a suburb that's home to many high-tech businesses, it's probably fair to assume that many households have Internet access. Such access to the Web isn't as likely to be found in poorer neighborhoods, however, and it is they who often have the more serious fire and life-safety problems.

It's a good idea for members of the department to have occasional visibility at meetings of elected bodies. While chiefs and staff officers of fire districts attend meetings of their own boards of directors, officers of municipal departments often don't have much contact with their city councils. It doesn't hurt to show an interest in the council's actitivies, as long as it isn't done in a way that suggests an effort to intimidate the council. It's also a good idea to create opportunities for elected officials to visit fire stations and do ride-alongs to learn firsthand about the work of fire agencies.

The agency I work for covers a large fire district, serving a suburban area that includes nine different cities. Each of the cities has been assigned an officer who serves as a liaison between the fire district and that particular municipality. The liaisons attend city council meetings from time to time, report periodically to the councils on fire district activities, have an occasional lunch with key city staff, and provide a first-name point of contact for city officials with fire-related issues or problems.

Many of the same principles can be used at the company and neighborhood level. In addition to having the department's identity boldly marked on apparatus, companies can also be marked with the name of the particular community or neighborhood they serve. Many of the engines and trucks in my department bear a slogan indicating that they proudly serve their particular city neighborhood. The Eugene Fire and EMS Department also places the names of individual city neighborhoods on apparatus. After years of designating its fire stations by number, Eugene Fire recently named its firehouses for the neighborhood they're located in. Thus, Station 3 is now known to the general public as University Station. These efforts get back to the sense of neighborhood identity discussed earlier in this chapter.

The same rules apply for community organizations. Many cities have formally constituted neighborhood associations. There should be no reason fire companies don't periodically attend the meetings of these organizations. In my experience, it isn't even necessary to have an item on the agenda. Just showing the flag as part-time neighborhood residents is enough.

At this point, I know that some readers are cringing. "You don't know the cast of characters I've got at Engine 13. There's no way I'd want them at a neighborhood meeting."

Okay, so it's a risk. A department in the midst of touchy contract negotiations or a major morale problem might think twice about encouraging its members to meet the public on company time, but refusing to do so as a matter of routine doesn't say much about the professionalism, hiring, and promotional practices of the department, either.

It's also important to recognize and provide positive reinforcement for those firefighters who do get involved in their neighborhoods. Many of the programs we've discussed in this chapter are things that, while they do increase public support for firefighters, also create additional work. Just as we recognize line-of-duty herosim, so should we recognize those members of our departments who go above and beyond the norm in community service. A few pizzas and commendations can go a long way toward sustaining positive community involvement. Tualatin Valley Fire and Rescue Chief Jeff Johnson has commissioned a special coffee mug that members can recieve only by performing exceptional acts of community or customer service.

Many departments have little choice as to whether or not they attend neighborhood association meetings—the meetings are held in their fire stations. A number of fire departments have adopted policies that new fire stations shall include community meeting rooms in their design. Such efforts capitalize on an advantage we have, in that, during an era of renewed neighborhood-level activism, fire departments by their nature have a decentralized infrastructure of stations, thereby allowing us to provide direct support to neighborhood affairs.

The future holds many challenges for the fire service: increased competition for public money, the impact of managed care on EMS programs, competition from the private sector, more mandates from other government agencies, the increasing complexity of codes and regulations, and a changing demographic and political environment. The fire departments that survive and thrive will be those that have truly learned to connect with their communities and deliver what their publics want in as efficient and cost-effective a manner as possible.

APPENDIX I

Model Public Information Policy and Procedure

Standard Operating Procedure: News Media Relations

I. Policy:

It is the policy of the department to cooperate with the news media whenever possible, within the guidelines of state public records law and the procedures set forth in this standard operating procedure. As a matter of policy, the department will communicate information to the fullest extent possible without compromising investigations or public safety.

II. Responsibility:

A. Alarm Office

1. On receipt of media inquiries regarding incidents in progress, the Alarm Office will confirm the basic nature and the reported location of the incident.
2. Alarm Office personnel must use caution in releasing information they have received from callers reporting incidents until such time as fire companies have arrived on scene and confirmed what has been report-

ed. Use special caution with unconfirmed information regarding fatalities and possible criminal activity. Information regarding deaths, serious injuries, and investigation status should be developed and released by the on-scene PIO or his designee.

3. The Alarm Office shall notify the on-call PIO of the following incidents:

- Multiple alarms.
- Mass-casualty incidents.
- Fire deaths.
- Major hazardous materials incidents.
- Major transportation accidents, such as aircraft and rail.
- Death or serious injury to an on-duty member.
- Any other incident having obvious interest to the news media.
- As requested by an on-duty chief or company officer.

B. Public Information Officer (PIO)

1. The PIO, or on-call designee, shall respond to all incidents as requested and report to the incident commander.
2. The PIO shall work with responding law enforcement agencies to establish a media area inside the perimeter established for the general public. This area should be selected to enable the media to get clear photographs and video of the event while ensuring their relative safety and noninterference with operations.
3. The PIO shall work closely with the incident commander in gathering information about the incident and determining what shall be released, pursuant to legal and policy guidelines.
4. The PIO shall establish and maintain a liaison with all on-scene media representatives, assisting them

with their newsgathering efforts while ensuring noninterference with department operations and preserving the integrity of any investigations.

5. The PIO shall provide periodic briefings to the media and should, if conditions allow, make the incident commander and/or department members involved in newsworthy actions available to the media for interviews.

6. At the conclusion of the incident, the PIO shall prepare a news release documenting the event and disseminate it to local news organizations. In cases involving sensitive matters or continuing investigations, the PIO shall review the release with the incident commander or fire investigators before release. Copies of the release shall be routed to the fire chief and the fire marshal.

7. In incidents of major significance, the PIO shall be responsible for notifying the fire chief of the event. On the chief's guidance, the PIO may also be charged with notifying the mayor and city council.

C. Fire Officers and Firefighters

1. The senior officer at the incident, or his designee, shall be the department's spokesperson in the absence of a PIO.

2. All personnel are cautioned not to give out any information relative to the cause of an incident unless such release has been authorized by the investigating authority.

3. All personnel are encouraged to cooperate with the media as much as possible. Members who speak to the media should limit the information they provide to that which they can confirm and of which they have firsthand knowledge. Even when members cannot accommodate a particular request from the media, they should strive to be as polite and courteous as possible.

4. Members should avoid releasing information of a medical or investigatory nature. Release of specific medical information regarding a named patient may constitute an invasion of personal privacy, and the release of information regarding the cause of an incident may compromise an investigation.
5. Members should not release the names of deceased or seriously injured persons.
6. Members may express personal opinions to the media as long as they clarify to the media that they are not official spokespersons for the department and that their observations are personal in nature.

D. Fire Chief

1. Only the fire chief may release the following information:

- Policy statements.
- Organizational changes.
- Information regarding disciplinary actions.
- Budget information.
- Staffing and deployment information.
- Statistical information.

APPENDIX II

Model Public Information Plan for Major Emergencies

City or County Emergency Operations Plan: Public Information Annex

I. Concept of Operations

A. In a major emergency necessitating the activation of the Emergency Operations Center (EOC), a public information function shall be established, including the appointment of a public information officer (PIO) and whatever additional staff may be necessary to fulfill the mission.

B. The PIO shall be located in the EOC or other specified location and shall have access to responsible elected officials, emergency managers, current information, and the news media.

C. The PIO shall serve as the official spokesperson and the media's single point of contact for the emergency management effort and will coordinate the release of information with elected and appointed officials overseeing the EOC and assume responsibility for the organization and operation of the public information function.

D. A continuing flow of emergency information and instructions shall be provided to the public and the news media.

E. Depending on the nature of the emergency, a rumor

control section may be established, as may dedicated phone lines to receive citizen inquiries.

F. The PIO shall coordinate with all other annex coordinators, liaison agencies, and PIOs from local, state, and federal agencies.

G. Ongoing public awareness and education programs will be conducted to make citizens aware of potential hazards and the mitigation, preparedness, response, and recovery activities associated with those hazards.

II. Organization and Responsibilities

A. Positions

1. At minimum, a single PIO shall be appointed. Depending on circumstances and need, additional staff in this unit may include an assistant PIO, a rumor control officer, and one or more field PIOs who shall perform news media liaison duties at incident scenes outside the EOC.
2. Communications shall be established and maintained among all members of this unit to ensure the consistency of information being released.

B. Required Tasks (by phase of emergency)

1. Mitigation Phase
   Develop and deliver a hazard awareness program covering such hazards as may be reasonably expected to occur within the jurisdiction and including mitigation procedures.
2. Preparedness Phase
   a. Develop ongoing relationships with all local and neighboring news media. Establish agreements with the media for dissemination of emergency public information and warnings as needed.
   b. Designate a media area adjacent to the EOC and any alternate sites.

c. Train staff on the role of the PIO and public information procedures in the event of an emergency.

d. Conduct public education programs on emergency response and recovery, warning signals, and evacuation routes.

e. Prepare emergency information packets for pre- and postdisaster use. Distribute pertinent materials to the media.

f. Brief news media personnel on emergency procedures for disseminating public information.

g. Develop a priority system for the release of public information—e.g., self-help, evacuation notices, shelter locations, and status of transportation systems and schools are of high priority; overall status of emergency is of medium priority; and human interest stories are of low priority until such time as critical information needs have been met.

h. Develop and maintain a file of prewritten news releases on such topics as self-help for various types of emergencies, preparedness information, evacuation advisories, and shelters.

i. Develop plans for reaching special populations, such as disabled and non-English-speaking persons.

j. Develop mechanisms for contacting all media during an emergency and rapidly releasing information.

k. Assign and train staff who will serve in public information functions during an emergency.

l. Participate in emergency operations drills and exercises.

3. Response Phase

a. On activation of the emergency plan, distribute information releases with as much accurate information as possible to all local media as quickly as possible. Follow up with regular updates.

b. If necessary, establish a rumor control function and publicize the phone numbers for citizen inquiries.

c. Serve as agency spokesperson, schedule news conferences, and respond to media inquiries.

d. Verify and confirm field reports.

e. Maintain a chronological record of emergency-related events.

f. Supervise the EOC media area.

g. Coordinate with media to facilitate news coverage without interfering with emergency operations.

h. Monitor news coverage of the emergency for accuracy and completeness of information.

i. Coordinate efforts with PIOs from other agencies and relief organizations to ensure the consistency of information being released to the public.

j. Disseminate information to special populations, as required.

4. Recovery Phase

a. Coordinate recovery information with other PIOs and release it as appropriate.

b. As directed by city or county officials, inform the public of those mitigation measures that were identified as a result of the disaster.

c. Assess the effectiveness of public information and education programs.

d. Prepare a post-action report detailing the activities of the public information unit. Include a chronological log of activities and recommendations for future events.

e. Archive all records from the event to assist with critiques and the processing of any claims.

f. Participate in post-event critique.

Oregon Bar/Press/ Broadcasters Joint Statement of Principles

Oregon's Bill of Rights provides for both fair trials and for freedom of the press. These rights are basic and unqualified. They aren't ends in themselves, but are necessary guarantors of freedom for the individual and the public's right to be informed. The necessity of preserving both the right to fair trial and the freedom to disseminate the news is of concern to responsible members of the legal and journalisitic professions, and is of equal concern to the public. At times, these two rights appear to be in conflict with one another.

In an effort to mitigate this conflict, the Oregon State Bar, the Oregon Newspaper Publishers Association, and the Oregon Association of Broadcasters have adopted the following statement of principles to keep the public fully informed without violating the rights of any individual.

1. The news media have the right and responsibility to print and to broadcast the truth.
2. However, the demands of accuracy and the objectivity in news reporting should be balanced with the demands of fair play. The public has the right to be informed. The accused has the right to be judged in an atmosphere free from undue prejudice.

3. Good taste should prevail in the selection, printing, and broadcasting of the news. Morbid or sensational details of criminal behavior should not be exploited.
4. The right of decision about the news rests with the editor or news director. In the exercise of judgment, he or she should consider that:
   a. an accused person is presumed innocent until proved guilty;
   b. readers and listeners are potential jurors;
   c. no person's reputation should be injured needlessly.
5. The public is entitled to know how justice is being administered. However, it is unprofessional for any lawyer to exploit any medium of public information to enhance his or her side of a pending case. It follows that the public prosecutor should avoid taking unfair advantage of his or her position as an important source of news; this shall not be construed to limit his or her obligation to make available information to which the public is entitled.

Guidelines for Disclosure and Reporting of Information on Criminal Proceedings

It is generally appropriate to disclose or report the following:

1. The arrested person's name, age, residence, employment, marital status, and similar biographical information.
2. The charge.
3. The amount of bail.
4. The identify of and biographical information concerning both the complaining party and the victim.
5. The identity of the investigating and arresting agency and the length of the investigation.
6. The circumstances of the arrest, including time, place, resistance, pursuit, and weapons used.

It is rarely appropriate to disclose for publication or to report prior to the trial the following:

1. The contents of any admission or confession, or the fact that an admission or confession has been made.
2. Opinions about an arrested person's character, guilt, or innocence.
3. Opinions concerning evidence or argument in the case.
4. Statements concerning anticipated testimony or the truthfulness of prospective witnesses.
5. The *results* of fingerprints, polygraph examinations, ballistics tests, or laboratory tests.
6. Precise descriptions of items seized or discovered during an investigation.
7. Prior criminal charges and convictions.

Special Statement

These guidelines are cautionary, not mandatory. They do not prohibit the release or publication of information needed to identify or aid in the capture of suspects or information required in the vital public interest after arrest. Neither do they proscribe publication of information that is already in the public domain.

Bibliography

There are any number of books, some available in local libraries, that can provide interested readers with more information on public and media relations. I highly recommend the following as good references to keep on hand:

American Bar Association. ABA *Standards for Criminal Justice: Fair Trial and Free Press, Third Edition.* An information manual for attorneys, news media, law enforcement agencies, and the courts, this booklet covers the conflicts described in Chapter Five and provides model guidelines for managing them. Write the American Bar Association, 750 N. Lake Shore Dr., Chicago, IL 60611 for ordering information, or call the ABA at (312) 988-5000. You can also order via the ABA's Website at www.aba.org.

Cutlip, Scott M., Allen H. Center, and Glen M. Broom. *Effective Public Relations.* Englewood Cliffs, N.J.: Prentice-Hall, Inc. Published in multiple editions, this is one of the standard texts for college-level public relations courses, covering all aspects of the craft. Used copies can often be found in secondhand bookstores.

Irvine, Robert B. *When You Are the Headline: Managing a Major News Story*. New York: Irwin Prof. Publishing, 1987. This is an advanced text for handling major events. The author presents a series of case studies, including the Challenger explosion, several commercial airline crashes, the Bhopal disaster, major medical events, and a mine disaster, to analyze the requirements for handling major news from a media relations perspective. The book is out of print but worth grabbing if you can locate a copy.

Jones, Clarence. *Winning With the News Media*. Now in its sixth edition, this is simply the best guide ever to working with the media, covering all aspects of media relations in an enlightening and entertaining way. Jones knows the business from both sides of the camera. Before becoming a news media consultant, crisis manager, and on-camera coach in 1984, he spent sixteen years as a newspaper reporter and fourteen years in television news. This book can be ordered on-line at www.winning-newsmedia.com.

Riha, Bob and Hunschuh, David. *National Media Guide for Emergency and Disaster Incidents*. National Press Photographers Association, 1995. Published by the national organization for photojournalists, this book should be on every PIO's bookshelf. It provides legal and procedural guidance for news photographers covering everything from wildland fires to undercover operations, and it includes the policies of a variety of federal and local agencies. Developed in cooperation with public safety officials, this book presents model standards for cooperation between news organizations and public safety agencies. Write the National Press Photographers Association, Inc., 3200 Croasdaile Drive, Suite 306, Durham, NC 27705, or call (800) 289-6772 for ordering information.

Selected Sources of Additional Training

Although sessions on public information are often found at fire service conferences, the following organizations offer structured coursework for PIOs.

The Emergency Management Institute (EMI), located in Emmitsburg, MD, and operated by the Federal Emergency Management Agency, offers basic and advanced PIO courses with an emphasis on public information in disasters and major emergencies. For more information, write EMI, 16825 South Seton Avenue, Emmitsburg, MD 21727.

The California Specialized Training Institute (CSTI), located near San Luis Obispo, CA, is operated by the Governor's Office of Emergency Services. CSTI offers three-week-long courses for public information officers in a three-year sequence and accepts non-California students at an out-of-state tuition rate. Write CSTI at P.O. Box 8123, San Luis Obispo, CA 93403-8123 for more information. CSTI can also be reached on-line at www.csti.org.

The National Information Officers Association (NIOA) is a professional organization for public safety

PIOs. The organization publishes a periodic newsletter and conducts an annual training conference, usually held in September. This conference typically includes "how-to" presentations and case studies of major incidents. For more information about the NIOA, write the organization at Box 10125, Knoxville, TN 37939. The NIOA can be reached on-line at nioa@aol.com.

The National Highway Traffic Safety Administration (NHTSA) has developed a training program called EMS PIER (Emergency Medical Services Public Information, Education, and Relations). The PIER course is a one-day training program that provides students with the basic skills and strategies needed to raise public awareness of emergency medical services and to initiate public education programs. Much of the course focuses on media, and the training includes the videotaping of practice interviews. NHTSA has developed a cadre of instructors who provide this training throughout the country. For information on how to connect with EMS PIER training in your state, contact: NHTSA, PIER Network, Emergency Medical Services Division, 400 Seventh Street SW, Room 5130 (NTS-14), Washington, D.C. 20590, or call (202) 366-4297.